DATE DUE

FE 10 97			
MR ~~98~~			
AP 20 98			
JE 16 00			
OG 28 04			

The Simple Science of Flight

The Simple Science of Flight

From Insects to Jumbo Jets

Henk Tennekes

The MIT Press
Cambridge, Massachusetts
London, England

© 1992 and 1996 Henk Tennekes

Originally published, under the title *De wetten van de vliegkunst: Over stijgen, dalen, vliegen en zweven,* by Aramith Uitgevers, Bloemendaal, The Netherlands. First English-language edition 1996.

Set in Melior and Helvetica by Wellington Graphics.
Printed and bound in the United States of America.

Library of Congress Cataloging-in-Publication Data
Tennekes, H. (Hendrik)
  [Wetten van de vliegkunst. English]
  The simple science of flight : from insects to jumbo jets / Henk
Tennekes. —1st English language ed.
  p. cm.
  Includes bibliographical references (p.) and index.
  ISBN 0–262–20105–4 (hc: alk. paper)
  1. Aerodynamics. 2. Flight. I. Title.
TL570.T46 1996
629.132′3—dc20
                                                                        95–35554
                                                                        CIP

A bird flies according to mathematical principles.
—Leonardo da Vinci, *Sul Volo degli Uccelli,* 1505

Contents

White-fronted goose (*Anser albifrons*): W = 17 N, S = 0.18 m^2, b = 1.40 m.

Introduction

This book is an act of revenge on the part of an assistant professor of aerospace engineering who dared to use flight calculations of ducks, geese, sparrows, and butterflies to entertain his class on aircraft performance. Two particularly humorless students complained to the head of the department: "We are studying aeronautical engineering because we are interested in aviation. Nowhere in the curriculum does it say that we have to study biology as well. Would you please ask Professor Tennekes to stick to the official syllabus?"

I was duly called in to explain my errant practices. The place was University Park, the main campus of Pennsylvania State University. The year was 1969.

"Henk, some of your students have complained," said my department head. "In your class you seem to have talked about geese and swans. I cannot condone that. Our profession—mine, and I trust yours, too—is a branch of engineering. Animals that flap their wings are none of our business. Please restrict yourself to airplane theory."

I was flabbergasted. It took me almost a minute before I managed to respond. "But that same theory also applies to the performance of birds. Isn't that a nice bonus?"

I have always been fascinated by the similarities between nature and technology. I learn by association, not by dissociation. Weren't a swan and a 747 designed with the same tender loving care? Notwithstanding their differences, they follow the same aerodynamic principles, and it is not that hard to explain how these principles work. Not to the minutest detail, though interested laypersons need not worry about the endless particulars with which scientists must cope.

Unfortunately, physicists and engineers have a habit that frightens many readers: they use harsh-looking formulas to calculate how large or how fast something should be, or how much energy it takes to get something done. Scientists have found that computing a few numbers makes it easier to gain insight into nature. Because most people avoid formulas like the plague, this creates a dilemma for the author of a book of science intended for a broad audience. Avoiding formulas altogether will make it seem that one has conjured the relevant numbers out of thin air—a kind of magic that will not contribute much to the readers' understanding. What, then, is the best way to introduce people to a field one loves?

The great marvels of science and technology are laid on thick in many popular science books. All too often the hidden message is: "Dear reader, you are but a layperson. You should have deep respect for the sophistication revealed to you by specialists, who are the ones who really understand the secrets of the universe, the building blocks of life, the fantastic blessings of computer technology, the great achievements of aerospace engineering."

I do not agree that one's respect for miracles is lessened by an attempt to understand them. On the contrary, one's sense of wonder can only grow as one's insight increases. After one has computed how large a swallow's wings should be, one's respect for the magnitude of the mystery that keeps the bird in the air can only be greater.

This is why I make no apology for including a few formulas and calculations in this book. I love doing simple calculations, and I hope that I can infect readers with this useful enthusiasm. When I read a newspaper story about trains that require 1 megajoule of energy per passenger-kilometer, and then notice at the breakfast table that 100 grams of peanut butter supply 2700 kilojoules, I cannot help calculating how much my train ticket would cost if the train ran on peanut butter rather than diesel oil or electricity.

In the same way, it doesn't help me much to know that a Boeing 747 uses about 12,000 liters of fuel an hour. Should I be impressed? I can understand the significance of that number only by comparing it to my car's fuel consumption. Numbers can only come to life through such comparisons. And if in the process an occasional formula is required, so be it. One doesn't have to be a specialist; in many cases anyone can do the simple sums needed to reach a reasoned conclusion.

Acknowledgments

With thanks to Fons Baede, Sylvia Barlag, Paul Bethge, Rob Brink-huijsen, Stanley Corrsin, Jim Deardorff, Ruth Engledow, Jeroen Gemke, Hein Haak, Hetty de Hoyer, Günther Können, Marlie van Laere, Bram Leutscher, Els Nijssen, Theo Opsteegh, Olga van der Pot, Robin Tennekes, Ruth Tennekes, John van der Torn, and Hans Wittenberg.

The Simple Science of Flight

Common tern (*Sterna hirundo*): W = 1.2 N, S = 0.056 m^2, b = 0.83 m.

Let us say that you are sitting in a jumbo jet, en route to some exotic destination. Half dozing, you happen to glance at the great wings that are carrying you through the stratosphere at a speed close to that of sound. The sight leads your mind to take wing, and you start sorting through the many forms of flight you have encountered: coots and swans on their long takeoff runs, seagulls floating alongside a ferry, kestrels hovering along a highway, gnats dancing in a forest at sunset. You find yourself wondering how much power a mallard needs for vertical takeoff, and how much fuel a hummingbird consumes. You remember the kites of your youth, and the paper airplane someone fashioned to disrupt a boring class. You recall seeing hang gliders and parawings over bare ski slopes, and ultralights on rural airstrips.

What about the wings on a Boeing 747? They have a surface area of 5500 square feet, and they can lift 800,000 pounds into the air—a "carrying capacity" of 145 pounds per square foot. Is that a lot? A 5 × 7-foot waterbed weighs 2000 pounds, and the 35 square feet of floor below it must carry 57 pounds per square foot—almost half the loading on the jet's wings. When you stand waiting for a bus, your 150 pounds are supported by shoes that press some 30 square inches (0.2 square feet) against the sidewalk. That amounts to 750 pounds per square foot—5 times the loading on the jet's wings. A woman in high heels achieves 140 pounds per square inch, which is 20,000 pounds per square foot.

From a magazine article you read on a past flight, you recall that a Boeing 747 burns 12,000 liters of kerosene per hour. A hummingbird consumes roughly its own weight in honey each day—about 4 percent of its body weight per hour. How does that compare to the 747? Midway on a long intercontinental flight, the plane weighs approximately 300 tons (300,000 kilograms; 660,000 pounds). The 12,000 liters of kerosene it burns each hour weigh about 10,000 kilograms (22,000 pounds), because the specific gravity of kerosene is about 0.8 kilogram per liter. This means that a 747 consumes roughly 3 percent of its own weight each hour.

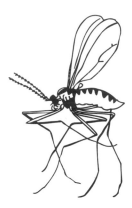

Crane fly (*Tipula lateralis*): $W = 3 \times 10^{-4}$ N, $S = 7.5 \times 10^{-5}$m^2, $b = 0.02$ m.

A hummingbird, however, is not designed to transport people. Perhaps a better comparison, then, is between the 747 and your car. At a speed of 560 miles per hour, the 747 uses 12,000 liters (3200 U.S. gallons) of fuel per hour—5.7 gallons per mile, or 0.18 mile per gallon. Your car may seem to do a lot better (perhaps 30 miles per gallon, or 0.033 gallon per mile), but the comparison is not fair. The 747 can seat up to 400 people, whereas your car has room for only four. What you should be comparing is fuel consumption per *passenger*-mile. A 747 with 350 people on board consumes 0.016 gallon per passenger-mile, no more than a car with two people in it. With all 400 seats occupied, a 747 consumes 0.014 gallon per passenger-mile. A fully loaded subcompact car consuming 0.025 gallon per mile (40 miles per gallon) manages 0.006 gallon per passenger-mile.

Ten times as fast as an automobile, at comparable fuel costs: no other vehicle can top that kind of performance. But birds perform comparable feats. The English house martin migrates to South Africa each fall, the American chimney swift winters in Peru, and the Arctic tern flies from pole to pole twice a year. Birds can afford to cover these enormous distances because flying is a relatively economical way to travel far.

Lift, Weight, and Speed

A good way to get started in understanding the basics of flight performance is to think about the weight a set of wings can support. This "carrying capacity" depends on a number of factors: wing size,

Sparrow hawk (*Accipiter nisus*): W = 2.5 N, S = 0.08 m^2, b = 0.75 m.

airspeed, air density, and the angle of the wings with respect to the direction of flight.

The role of wing size is straightforward: a wing's aerodynamic lift L is proportional to its surface area S. In practice, S is defined as the surface area measured from a full-scale photograph of the wings taken from above. The relation between L and S is simple enough: a wing twice as large can carry twice as much weight.

The relation between lift and airspeed is less straightforward. If we call the airspeed V (for "velocity") and the air density d, then the mass flow of air around the wings is proportional to d times V (written as dV). According to a version of Newton's Second Law of Motion that is explained in detail in chapter 4, the force generated by the airflow around the wings is proportional to V times dV. (That product is expressed as DV^2.) If one flies twice as fast, with the same wings at the same angle in the air stream, the lift is 4 times as great. And if one flies at an altitude of 12 kilometers (39,000 feet), where the air density is only one-fourth its value at sea level, one must fly twice as fast to sustain the same weight.

Now let's look at the angle between the wing and the airflow it encounters, known as the "angle of attack." It is easy to experiment with angle of attack: just stick your hand out of the window of a car moving at speed. Keep your hand level and you feel only air resistance, but turn your wrist and your hand wants to move up or down. You are then generating aerodynamic lift. But as you increase the angle of your hand with the airstream, you start generating more resistance while losing much of the lift.

Birds and aircraft can change the angle of attack of their wings to fit the circumstances. They fly nose up, with a high angle of attack, when they have to fly slowly; they fly nose down when speeding. But all use about the same angle of attack in long-distance cruising; 6° is a reasonable average.

Since wings have to support the weight of an airplane or a bird against the force of gravity, the lift L must equal the weight W. The lift is proportional to the wing area S and to dV^2, and so is the weight:

$$W = 0.3dV^2S. \tag{1}$$

(Where does the 0.3 come from? It is related to the angle of attack in long-distance flight, for which the average value of 6° has been adopted.)

We must make sure we aren't violating the rules of physics when we use equation 1. We must give clear and mutually consistent definitions for the units in which d, V, and S are expressed. (Clearly the numbers would look different if velocities were given in miles per minute rather than in millimeters per hour.) The best way to ensure consistency is to use the metric system of physical units, expressing S in square meters, V in meters per second, and d (air density) in kilograms per cubic meter. The rules of physics then require that the weight W in equation 1 be given in kilogram-meters per second squared. This frequently used unit is known as the *newton,* after Sir Isaac Newton (1642–1727), the founder of classical

Dragonfly (*Aeschna cyanea*): W = 0.01 N, S = 0.0018 m^2, b = 0.1 m.

Razorbill (*Alca torda*): W = 8 N, S = 0.038 m^2, b = 0.68 m.

mechanics. A newton is slightly more than 100 grams (3.6 ounces). A North American robin weighs about 1 newton, a common tern a little bit more, a starling a little bit less. Since there are roughly 10 newtons to a kilogram, a 70-kilogram (154-pound) person weighs about 700 newtons.

If we respect the rules, we can play with equation 1 in whatever way we want. For example, a Boeing 747 has a wing area of 5500 square feet (511 square meters) and flies at a speed of 560 miles per hour (900 kilometers per hour; 250 meters per second) at an altitude of 12 kilometers (40,000 feet), where the air density is only one-fourth its sea-level value of 1.25 kilograms per cubic meter. Using d = 0.3125 kilograms per cubic meter, V = 250 meters per second, and S = 511 square meters, we calculate from equation 1 that W must equal 2,990,000 newtons. Because a newton is about 100 grams, this corresponds to roughly 300,000 kilograms, or 300 tons. That is indeed the weight of a 747 at the midpoint of an intercontinental flight. At takeoff it is considerably heavier (the maximum takeoff weight of a 747-400 is 394 tons), but it burns 10 tons of kerosene per hour.

Equation 1 can be used in several ways. Consider a house sparrow. It weighs about an ounce (0.3 newton), flies close to the ground (so that we can use the sea-level value of d, 1.25 kilograms per cubic meter), and has a cruising speed of 10 meters per second (22 miles per hour). We can use equation 1 to find that the sparrow needs a wing area of 0.01 square meter, or 100 square centimeters. That's 20 centimeters from wingtip to wingtip, with an average width of 5 centimeters. Or we can use the same equation in designing a hang glider. Taken together, the pilot and the wing weigh about 1000 newtons (100 kilograms; 220 pounds). So if you want to fly as fast as a sparrow (20 miles per hour), you need wings with a surface area of 33 square meters. On the other hand, if you want to fly at half the speed of a sparrow, your wing area must be more than 100 square meters (more than 1000 square feet).

Wing Loading

To make equation 1 easier to work with, let us replace the variable d (air density) with its sea-level value: 1.25 kilograms per cubic meter. This should not make any difference to birds, which fly fairly close to the ground. For aircraft flying at higher altitudes, we will have to correct for the density difference or return to equation 1; we can worry about that detail when it becomes necessary. Another improvement in equation 1 is to divide both sides by the wing area S. The net result of these two changes is

$$\frac{W}{S} = 0.38V^2. \tag{2}$$

This formula tells us that the greater a bird's "wing loading" W/S, the faster the bird must fly. Within the approximations we are using here, sea-level cruising speed depends on wing loading only. No other quantity is involved. This is the principal advantage of equation 2.

The predecessor of the Fokker 50 was the Fokker Friendship, with a weight of 19 tons (190,000 newtons) and a wing area of 70 square meters. Its wing loading was 2700 newtons per square meter, good for a sea-level cruising speed of 85 meters per second (190 miles per hour). The wing loading of a Boeing 747 is about 7000 newtons per square meter, and it must fly a lot faster to remain airborne. The wing loading of a sparrow is only 38 newtons per square meter, corresponding to a cruising speed of 10 meters per second (22 miles per

Table 1 Weight, wing area, wing loading, and airspeeds for various seabirds, with *W* given in newtons (10 newtons equal 1 kilogram, roughly), *S* in square meters, and *V* in meters per second and miles per hour. The values of *W* and *S* are based on measurements; those for *V* were calculated from equation 2. In general, larger birds have to fly faster.

| | | | | *V* | |
	W	*S*	*W/S*	m/sec	mph
Common tern	1.15	0.050	23	7.8	18
Dove prion	1.70	0.046	37	9.9	22
Black-headed gull	2.30	0.075	31	9.0	20
Black skimmer	3.00	0.089	34	9.4	21
Common gull	3.67	0.115	32	9.2	21
Kittiwake	3.90	0.101	39	10.1	23
Royal tern	4.70	0.108	44	10.7	24
Fulmar	8.20	0.124	66	13.2	30
Herring gull	9.40	0.181	52	11.7	26
Great skua	13.5	0.214	63	12.9	29
Great black-backed gull	19.2	0.272	71	13.6	31
Sooty albatross	28.0	0.340	82	14.7	33
Black-browed albatross	38.0	0.360	106	16.7	38
Wandering albatross	87.0	0.620	140	19.2	43

hour). From these numbers one gets the impression that wing loading might be related to size. If larger birds have higher wing loadings, it is no coincidence that a Boeing 747 flies much faster than a sparrow.

Our understanding of the laws of nature is due in large part to people who have been driven by the urge to investigate such questions. One person in particular deserves to be mentioned here: Crawford H. Greenewalt, a chemical engineer who was chairman of the board of Du Pont and a longtime associate of the Smithsonian Institution. For many years Greenewalt's chief hobby was collecting data on the weights and wing areas of birds and flying insects. Hummingbirds were his favorites, and he carried out many strobe-light experiments to measure their wing-beat frequencies.

Some of the data collected by Greenewalt and other investigators are listed in table 1. For the sake of clarity, the selection is restricted to seabirds: terns, gulls, skuas, and albatrosses. Looking at table 1, we find that wing loading and cruising speed generally increase as birds become heavier. But the rate at which this happens is not spectacular. A wandering albatross is 74 times as heavy as a common tern, but its wing loading is only 6 times that of its small cousin, and it flies only 2.5 times as fast (equation 2). In terms of weight, the wing loading isn't terribly progressive.

To improve our perception of what is happening, let us plot the weights and wing loadings of table 1 in a proportional or "double-logarithmic" diagram, which preserves the relative proportions between numbers. In a proportional diagram a particular ratio (a two-fold increase, say) is always represented as the same distance, no matter where the data points are located. Four is 2 times 2, and 100 is 2 times 50; in a proportional diagram the distance between 2 and 4 is equal to the distance between 50 and 100. Take your time to check this out in figure 1.

The steeply ascending line in figure 1 suggests that there must be a simple relation between size and wing loading. There are deviations from the line, of course; for example, the fulmar has a rather high wing loading for its weight. But before we look at the exceptions, we have to explain the rule.

All gulls and their relatives look more or less alike, with long, slender wings, pointed wingtips, and a beautifully streamlined body with a short neck and tail; however, they vary considerably in size. Now compare two types of gull, one having twice the wingspan of the other. If the larger of the two is a scaled-up version of its smaller cousin, its wings are not only twice as long but also twice as wide, making its wing area 4 times as large. The same holds for weight. Because weight goes as length times width times height, the weight of the larger gull is 8 times that of its smaller cousin. Eight times as heavy, with a wing area 4 times as large, a bird with a wingspan twice that of its smaller cousin has twice the wing loading, though it is 8 times as heavy. And according to equation 2 it has to fly 40 percent faster (the square root of 2 is about 1.4).

It is useful to write this down in an equation. If the wingspan (the distance from wingtip to wingtip with wings fully outstretched) is called b, the wing area is proportional to b^2 and the weight is proportional to b^3. The wing loading, W/S, then is proportional to b. But b itself is proportional to the cube root of W. In this way we obtain the scale relationship

$$\frac{W}{S} = c^3\sqrt{W} \, . \tag{3}$$

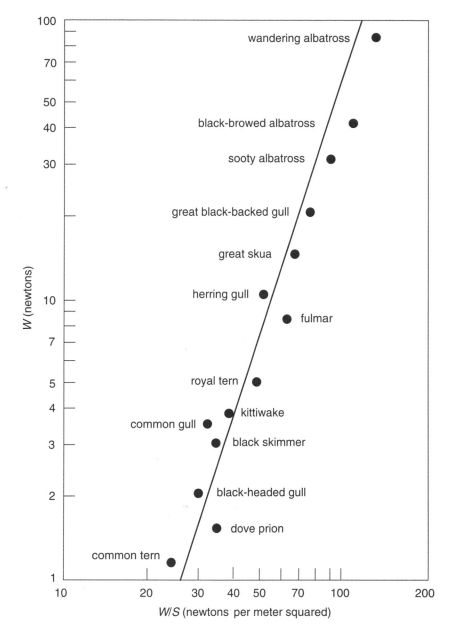

Figure 1 The relation between weight and wing loading represented in a proportional diagram. When the weight increases by a factor of 100, the value of *W/S* increases by a factor of 5 and the airspeed by a factor of more than 2.

Herring gull (*Larus argentatus*): W = 9.4 N, S = 0.18 m^2, b = 1.4 m.

Strictly speaking, this formula holds only for birds that are "scale models" of one another. The steeply ascending line in figure 1 corresponds to equation 3, the coefficient having been determined empirically. For the seabirds in figure 1, c is 25: at a weight of 1 newton, the wing loading is 25 newtons per square meter.

The scale relation, equation 3, is universally applicable whenever weights and supporting surfaces or cross-sectional areas are involved. Galileo Galilei (1564–1642) wrote the first scientific treatise on this subject, asking himself why elephants have such thick legs and similar questions. The answer is that the larger an animal gets, the more crucial the design of the legs becomes. The stress on leg bones increases as the cube root of weight; for this reason, a land animal much larger than an elephant is not a feasible proposition. This is the same problem that engineers face when they design bridges, skyscrapers, or even stage curtains, which would give way under their own weight were they not reinforced by steel cables. Another good example is that of walking barefoot on a stony beach. For adults walking on gravel can be an uncomfortable experience, but little children may frolic around them unconcerned. A father who is twice as tall and 8 times as heavy as his 8-year-old daughter must support himself on feet whose surface area is only 4 times that of her feet. Thus, his "foot loading" is twice hers. No wonder he seems to be walking on hot coals.

The scale relation given in equation 3 is not a hard and fast rule. Most birds are not exact "scale models" of others, and we must also allow some latitude for deviations to fit designers' visions. On the other hand, designers are confronted by tough technical problems whenever they deviate too far. The margins permitted by the laws of scaling are finite.

The Great Flight Diagram

Thanks to the dedicated work of Crawford Greenewalt and other enthusiasts, and assisted by the great airplane encyclopedia *Jane's All the World's Aircraft,* we can now put everything that can fly together in a single proportional diagram (figure 2). The results are impressive: *12* times a tenfold increase in weight, *4* times a tenfold increase in wing loading, and *2* times a tenfold increase in cruising speed! Very few phenomena in nature cover so wide a range. At the very bottom of the graph we find the common fruit fly, *Drosophila melanogaster,* weighing no more than 7×10^{-6} newton (less than a grain of sugar), with a wing area of just over 2 square millimeters. At the top is the Boeing 747, which weighs in at 3.5×10^{-6} newtons, 500 billion times as much as a fruit fly. The 747's wings, with an area of 511 square meters, are 250 million times as large. Yet, despite these enormous differences, a 747 flies only 100 times as fast as a fruit fly.

Allow yourself time to study figure 2 carefully. It is loaded with information. The ascending diagonal running from bottom left to top right is the scale relation of equation 3. The constant c has been set equal to 47, almost twice the value used for seabirds. The vertical line marks a cruising speed of 10 meters per second, corresponding to 22 miles per hour and to force 5 on the Beaufort scale used by marine meteorologists. Birds that fly slower than this (those to the left of the vertical line) may not be able to return to their nest in a strong wind. (To return home in a headwind, a bird must be able to fly faster than the rate at which the wind sets it back.)

Deviations from the rule can be seen both to the left and to the right of the diagonal representing the scale relation of equation 3. The diagonal acts as a reference, a "trend line," a standard against which individual designs can be evaluated. Let's start with the birds and airplanes that follow the trend—the run-of-the-mill designs, the com-

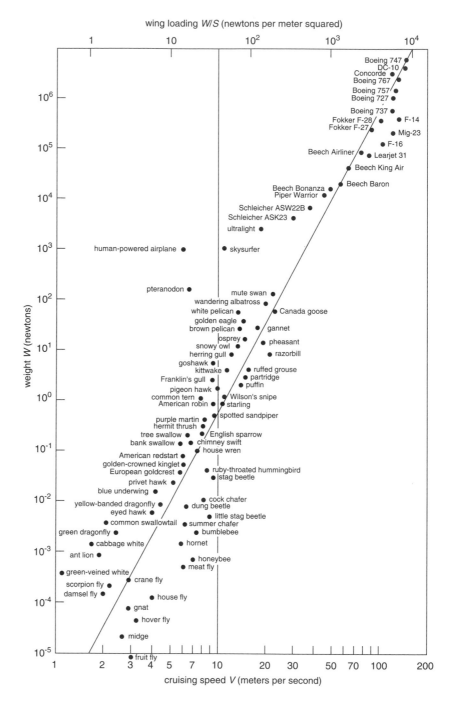

Figure 2 The Great Flight Diagram. The scale for cruising speed (horizontal axis) is based on equation 2. The vertical line represents 10 meters per second (22 miles per hour).

Storm petrel (*Hydrobates pelagicus*): W = 0.17 N, S = 0.01 m^2, b = 0.33 m.

monplace, homely types found on or near the diagonal. The starling
is a good example. A thrush-size European blackbird, 100 of which
were released in 1890 in New York's Central Park, it has become a
most successful immigrant. With a weight of 0.8 newton (80 grams, a
little over 3 ounces) and a wing loading of 40 newtons per square
meter, the starling is clearly an ordinary bird and does not have to
meet any special performance criteria. But the Boeing 747 also
follows the trend. In its weight class the 747 is a perfectly ordinary
"bird," with ordinary wings and a middle-of-the-road wing loading.
The only thing that is special about the 747 is its weight—not many
birds weigh 350 tons.

Deviations from the trend line may be necessary when special
requirements are included in the design specifications. The 747's
little brother, the 737, weighs only 50 tons (5×10^5 newtons), one-
seventh the weight of a 747. If the 737 had been designed as a scale
model of the 747, its wing loading would have been almost half that
of the larger plane (the cube root of 7 is 1.9). And according to
equation 2 its cruising speed would have been only 73 percent of its
big brother's: not 560 but only 410 miles per hour. This would have
been a real problem in the dense air traffic above Europe and North
America, where backups are much easier to avoid if all planes fly at
approximately the same speed. To make it almost as fast as the 747,
the 737 was given undersize wings. Its wing loading is higher than
those of ordinary planes of the same weight class, and it is therefore

Barn swallow (*Hirunda rustica*): W = 0.2 N, S = 0.013 m^2, b = 0.33 m.

located to the right of the trend line in figure 2. (With a cruising speed 60 mph less than that of the 747, the 737 is still a bit of a nuisance in dense traffic.)

To the left of the diagonal in figure 2 are several birds whose wing loadings are below average in their weight class. With their disproportionately large wings, these birds have low airspeeds for their size. Also in this group is the pteranodon, the largest of the flying reptiles that lived in the Cretaceous era. Weighing 170 newtons (37 pounds), it was almost twice as heavy as a mute swan or a California condor. The pteranodon had a wingspan of 23 feet (7 meters) and a wing area of 108 square feet (10 square meters)—comparable to a glider. Its wing loading was 17 newtons per square meter, about one-tenth that of a swan but comparable to that of a swallow. A pteranodon spent its life soaring above the cliffs along the shoreline, since it was not nearly strong enough for continuous flapping flight. Its airspeed was about 7 meters per second (16 miles per hour)—not fast for an airborne animal that must return to its roost in a maritime climate. However, there were no polar ice caps during the Cretaceous era, and there was less of a temperature difference between the equator and the poles than today. As a result there was much less wind.

After centuries of experimentation, humans finally managed to fly under their own power. Achieving that miracle required feather-light machines with extremely large wings. The only way to reduce the power requirement to a level that humans could attain was to reduce the airspeed to an absolute minimum. Humans pedaling through the air on gossamer wings at 11 miles per hour are the real mavericks in the Great Flight Diagram.

The more you study figure 2, the more you see. What about the Concorde? Isn't it supposed to fly at about 1300 miles per hour? How come it doesn't have higher wing loading and therefore smaller wings? The answer is that the Concorde suffers from conflicting design specifications. Small wings suffice at high speeds, but large wings are needed for taking off and landing at speeds comparable to those of other airliners. If it could not match such speeds, the Concorde would require special, longer runways. The plane's predicament is that it has to drag oversize wings along when cruising in the stratosphere at twice the speed of sound. No wonder its fuel consumption is outrageous. (Buy a Concorde ticket and you'll understand.) Adjustable wings would be a perfect solution. Many birds manage to adjust their wings effortlessly, but for most aircraft the massive hinges needed would be far too heavy and cumbersome. Only the Air Force can afford the luxury of hinged wings.

Since we are talking about adjustable wings, what about the various swifts, swallows, and martins in the Great Flight Diagram? They are all found on the left of the trend line. For their weight, they all have rather large wings and fly relatively slowly. There must be something wrong here. Swifts did not get their name for nothing.

Swifts and their relatives can fly very slowly, when they have to, by spreading their wings wide. When they want to fly faster, they can fold their wings. The elegance of their streamlining does not suffer when they reduce their wing area, but the wing loading increases, and with it the cruising speed. Are they poking fun at the laws of nature? According to equation 2, a bird cannot alter its speed at will if it wants to fly economically, once blessed with a particular set of wings. The cruising speed is controlled by the wing loading: $W/S = 0.38V^2$. But if S can be changed to fit the circumstances, this problem vanishes: the cruising speed then changes, too. All birds do this to some extent, though not always with the grace and sophistication of swifts and swallows.

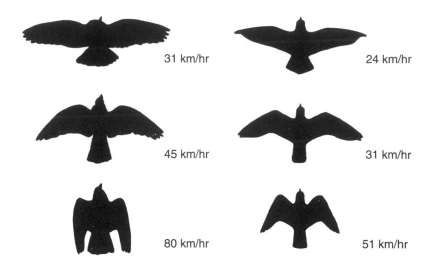

31 km/hr   24 km/hr

45 km/hr   31 km/hr

80 km/hr   51 km/hr

Figure 3 Birds progressively fold their wings as their speed increases. On the left is a pigeon, on the right a falcon. At high speeds, fully spread wings generate unnecessary drag; this can be avoided by reducing the wing area.

Figure 4 An ibis on final approach, with all flying feathers spread (including those of the tail) and with landing gear fully extended.

As figure 3 shows, pigeons and falcons fold their wings as their speeds increase. As figure 4 shows, it is easy to spot a bird preparing to land: trying to fly as slow as possible, it spreads its wings wide, separating the primary quills if it can. Airplanes do the same thing, with various flaps and slats that are partially extended at takeoff and fully extended at landing. Airplanes and birds alike minimize their landing speed to reduce the length of runway required or the risk of stumbling over their own feet.

The Boeing 747 is represented in figure 2 as having a wing loading of 7000 newtons per square meter and a cruising speed of 136 meters per second. But 136 meters per second is 300 miles per hour, roughly half the 747's actual cruising speed. What has gone wrong here? The problem is that in figure 2 the lower air density at cruising altitudes has been ignored. Since the air density at 39,000 feet is only one-fourth the density at sea level, the high-altitude cruising speed is twice the cruising speed near the earth's surface. Figure 2 gives the speed at sea level; table 6 (in chapter 6) gives the necessary conversion factors.

A curious feature of figure 2 is the continuity between the largest insects and the smallest birds. The largest of the European beetles, the stag beetle *Lucanus cervullus,* weighs in at 3 grams, about the same as a sugarcube or a fat hazelnut. The smallest European bird, the goldcrest, weighs just a little more: 4 grams. The wing loadings of large insects do not differ much from those of small birds, either. This is no minor observation. In theory, conditions may be imagined in which the largest beetle exceeds the smallest bird in size, or a wide gap exists between the largest flying insects and the smallest birds. Such a gap does exist between the largest birds and the smallest airplanes, after all. And there are substantial construction differences, too. The exoskeletons of insects are made up of load-bearing

Cockchafer (*Melolontha vulgaris*): W = 0.01 N, S = 0.0004 m^2, b = 0.06 m.

Common tern (*Sterna hirundo*): W = 1.2 N, S = 0.056 m², b = 0.83 m.

An insurance-company physician may put you on a fancy exercise machine, turn up the dial to 150 watts, and wait impassively for you to lose your breath. Having achieved that, the physician turns the dial back a little to determine the load at which your heart stabilizes at 120 beats a minute. The Consumers Union does much the same with cars, driving them hard on a test bench to measure engine power, fuel consumption, and emissions. Aeronautical engineers do their testing in wind tunnels, where powerful fans simulate the high speeds needed to put aircraft models through their paces. The use of sophisticated scale models, often complete with model engines, is important. Aircraft companies need to know the flying properties of their new airplanes in great detail before they send someone off on a first test flight.

Wind tunnels come in all shapes and sizes. The simplest is a straight pipe with an adjustable ventilator at one end, but most are much more elaborate and expensive. Some, like the supersonic wind tunnels used to test models of jet fighters and of the Space Shuttle, require enormous amounts of power. The very largest have working cross sections several meters wide and high. Wind tunnels are used not only to test airplanes but also to investigate how air flows around full-scale cars and even to do environmental impact studies of industrial installations.

But Vance Tucker's wind tunnel beats them all. Tucker, a zoologist at Duke University, spent many years investigating the flight performance of birds. He started, many years ago, by training a budgerigar to fly in a specially built wind tunnel. In order to measure its oxygen consumption, he fitted the little bird with an oxygen mask. That way he could get the data he needed to calculate how much energy the bird used at various airspeeds (ranging from 10 to 30 miles per hour) under various conditions, in horizontal flight and during 5° ascents and descents (figure 5). The budgy, not a particularly proficient flyer to begin with, no doubt would have preferred to spend its time preening for its mate.

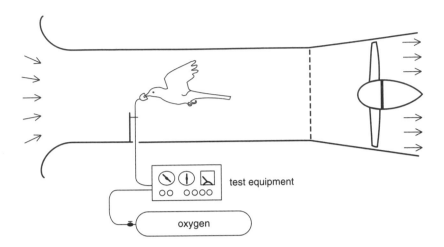

Figure 5 Vance Tucker's budgerigar in the wind tunnel.

Once he had his oxygen-consumption data, Tucker still needed to convert it into useful form. First he had to calculate the energy consumed during each flight. Next he had to subtract the energy spent maintaining the bird's metabolism (which would not be available for flight propulsion). The basal metabolic rate in birds is roughly 20 watts per kilogram of body weight, 10 times the rate in humans. Tucker's budgerigar weighed 35 grams, so its body used 0.7 watt to sustain itself. Finally, since the efficiency of the conversion from metabolic to mechanical energy is only about 25 percent, the net propulsive power is one-fourth the metabolic cost of flying. The net mechanical power of the flight muscles is plotted in figure 6.

Figure 6 deserves close attention, since it raises a number of questions about the energy requirements of aerial locomotion. But the most striking feature of figure 6 is that slow flight is uneconomical. It is easy enough to understand that the faster you travel the more power you need; riding a bike or driving a car will have taught you that. Strange as it may sound, birds also need a lot of power if they want to fly slowly. For this reason the power required for flapping flight has a minimum in the middle of the speed range. (Figure 6 shows this.) In horizontal flight, the most economical speed for the budgy was slightly more than 8 meters per second (18 miles per hour). At that speed it required 0.75 watt, equivalent to a thousandth of a horsepower, to remain airborne. At lower and higher speeds more power was needed.

Low-speed flight is uneconomical because birds and airplanes have to push the air surrounding them downward in order to stay

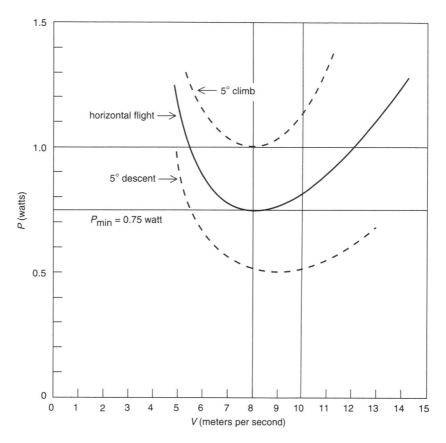

Figure 6 The flight performance of Tucker's budgerigar. In horizontal flight the most economical conditions were obtained at about 8 meters per second, with the bird supplying 0.75 watt of mechanical power. In climbing flight more power is needed; in descending flight less.

airborne. Relatively little air flows around the wings when the airspeed is low. To sustain its own weight, a bird or an aircraft must give that small quantity of air a powerful impulse. That requires a lot of energy. When a bird is flying fast, though, a lot of air crosses its wings; only a little push is needed to remain airborne then, and that requires much less energy. These issues will be taken up again in chapter 4.

Tucker's budgy apparently could not fly slower than 5 meters per second (11 miles per hour). At that speed, flapping its wings at full power, its flight muscles generated about 1.3 watt, evidently its maximum continuous power output. Let's do a few calculations to see if these numbers make sense.

The pectoral muscles of a bird account for about 20 percent of its body weight; in continuous flapping flight, a bird can generate about

100 watts of mechanical power per kilogram of muscle mass. A human can manage only about 10 watts per kilogram of muscle mass. No wonder it took us so long to learn to fly under our own power.

Since the pectoral muscles of a 35-gram budgerigar weigh about 7 grams, the corresponding continuous power output should be about 0.7 watt, corresponding nicely to the minimum in figure 6. The maximum continuous power of flight muscles is roughly twice the normal rate, or 200 watts per kilogram of muscle mass. For the budgy this works out to 1.4 watts, pretty close to the experimental data in figure 6. When the rules of thumb you are using agree with the results of measurements, you know are on the right track.

Energy, Work, and Power

We have made a terrible mess of simple physical concepts in ordinary life. We treat force, power, and energy as if they were interchangeable. Straightforward, unequivocal definitions of these quantities have been around for 200 years, and yet we continue to be confused. The scientific definition of the word "power" is "rate of work," nothing more and nothing less. In everyday language, of course, the word "power" has many other connotations. My Random House Dictionary refers to capabilities and capacities, to political strength (incidentally, wouldn't "strength" be a force?), to control and command, and so forth. Half a column goes by before it gets around to stating: "in physics, the time rate of doing work." Indeed, the majority of concepts associated with the word "power" tend to confuse rather than enlighten. The same holds for the words "force" and "energy." The Danes speak of nuclear "force" when they mean energy. Their bumper stickers proclaim "Atomkraft—nei tak" ("Atomic en-

White stork (*Ciconia alba*): W = 34 N, S = 0.5 m^2, b = 2m.

ergy—no, thanks"). The Dutch and the Germans use the equivalent of "horseforce" instead of "horsepower." Psychologists—whatever their native tongue—refer to mental energy as though it satisfied some mechanistic conservation law.

When we are trying to describe nature, we need clear and unambiguous definitions. Let us agree, then, that a force is the intensity with which I push or pull at an object, whether it starts to move or not. Exerting as much force as I please, I perform no work unless the object is displaced in the same direction. Sideways displacements do not count. Work is performed in proportion to the distance the object is displaced. Work is a form of energy, and the time rate at which it is consumed or supplied is called power.

When calculating the amount of work performed, we must consider both the force applied and the distance the object has moved. What is more straightforward than that work equals force times distance? That indeed is the accepted definition. The corresponding definition of power is equally straightforward: power is a rate of doing work, or a certain amount of energy expended per second. Since energy equals force times distance, power must be force times distance per second. But distance traveled per second is what we call speed. Therefore, power equals force times speed.

Before continuing with the flight performance of birds, we need to define the proper units for work (energy) and power. In chapter 1 we settled on the newton (102 grams) as our force unit. Defining the units for work and power is simple. Work equals force times distance, so in the metric system it must be calculated in newton-meters. This is a unit in itself, called the joule after the British physicist James Joule (1818–1889), who performed a brilliant energy-conversion experiment in 1845. Power equals energy per second and so must be computed in joules per second. But because a joule equals a newton-meter, power may also be calculated in newton-meters per second; that comes to the same thing. The unit for power also has a name of its own: the watt, after James Watt (1736–1819), the Scottish inventor of the steam engine. We used watts earlier in this chapter when discussing the power requirements of Tucker's budgerigar. (James Watt used another power unit, for which he coined the term "horsepower." He needed a word that would kindle the imagination of coal mine directors who were in the market for replacing their pit ponies with one of his steam engines. One horsepower is equivalent to 750 watts, and because there are 1000 watts in a kilowatt a horsepower equals 0.75 kilowatt.)

Table 2 The metric units for energy (force times distance), power (force times speed), and force (energy per unit distance).

Energy
1 joule = 1 newton-meter
       = 1 watt-second

Power
1 watt = 1 joule per second

Force
1 newton = 1 joule per meter

The quickest way to get used to a new set of concepts is to play with them. You know from experience that it takes energy to walk uphill, because you have to lift your own weight against the pull of gravity. How much energy does it take to climb one flight of stairs? Let's assume that you weigh 70 kilograms (about 154 pounds, or 700 newtons) and that the vertical distance between floors is 3 meters (about 10 feet). Since energy equals force times distance, it takes 2100 joules to climb from floor to floor. Is that a lot? Not at all. A gram of peanut butter contains 27 kilojoules (27,000 joules) of nutritional energy. If your body converts 20 percent of that to useful work, it will have 5400 joules to expend. This means that you can climb more than two flights of stairs on a single gram of peanut butter. Walking up and down flights of stairs at the office is not an efficient way to lose weight.

The *power* you need to walk up a flight of stairs (that is, the *rate* at which energy has to be supplied) is not negligible, however. If your

Ruby-throated hummingbird (*Archilochus colubris*): W = 0.03 N, S = 0.0012 m^2, b = 0.09 m.

Black-headed gull (*Larus ridibundus*): W = 2.3 N, S = 0.075 m², b = 0.90 m.

vertical speed on the stairs is 0.5 meter per second, only half the speed typical of a leisurely hiking trip, the climbing power you need (force times speed) is 700 newtons times 0.5 meter per second, which equals 350 watts. Only a professional athlete can maintain such a rate for more than a minute or so. A healthy amateur can maintain a power output of 200 watts for less than an hour; a professional athlete can maintain this rate for several hours.

The performance of Tucker's budgerigar was measured both in horizontal flight and in ascents of 5° (which corresponds to a slope of 8.7 percent). Flying at the minimum-power speed of 8 meters per second, the budgy achieved a rate of ascent of 0.7 meter per second. Now we can repeat the calculation we did a minute ago. Power equals force times speed; the force we are talking about here is the force needed to lift the bird's weight upward. The budgy weighed 0.35 newton (35 grams, a little more than an ounce), and the climbing power required was 0.35 newton times 0.7 meter per second, which is about 0.25 watt. The power needed to sustain horizontal flight at a speed of 8 meters per second was 0.75 watt. In climbing flight an additional 0.25 watt was needed, for a total of 1 watt. Figure 6 confirms this calculation.

A fourth of a watt is not even enough to light a single bulb on a Christmas tree. A 350-ton Boeing 747's rate of climb immediately after takeoff is about 15 meters per second, or 3000 feet per minute in aviation parlance. Quite apart from the power needed to remain airborne, the four jet engines of a 747 then must supply 50 million watts. Just imagine: 50 megawatts, equivalent to 17 top-of-the-line diesel locomotives producing 3000 kilowatts (4000 horsepower) each.

When you climb you need to exert yourself to overcome the force of gravity, but when you descend gravity does part of the work. To descend at a glide slope of 5° at 8 meters per second, Tucker's budgy needed only 0.5 watt of power: 0.75 watt to remain airborne, minus the 0.25 watt supplied courtesy of gravity. Taking this a bit further, we can easily work out the glide angle at which a budgy no longer needs to flap its wings: about 15°, corresponding to a glide slope of about 26 percent. Thus, for every meter of altitude it loses, a budgerigar travels about 4 meters forward. Is that a decent figure? A seagull can easily reach 10 meters of forward travel for every meter of altitude it loses.

Power equals force times speed. The power required in horizontal flight must be equal to some unknown force times the forward speed. But what is this force? It is the propulsive force, or thrust, T. To fly at a constant speed, a bird or a plane must develop just enough thrust to overcome aerodynamic drag D (see chapter 4). Since the two forces must be equal at constant speed, we shall just talk about the drag from now on. Drag times speed equals the power required, P, so drag equals power divided by speed. In notation:

$$P = DV \tag{4}$$

and

$$D = \frac{P}{V}. \tag{5}$$

It is quite simple to convert figure 6 into a graph depicting drag versus speed. For every point on the power curve for horizontal flight we can determine P along the ordinate and V along the abscissa. Using equation 5, we can then compute the value of D at each speed. The result is shown in figure 7.

Figure 7 demonstrates even more clearly than figure 6 that low-speed flight is extremely uneconomical. The minimum drag of a

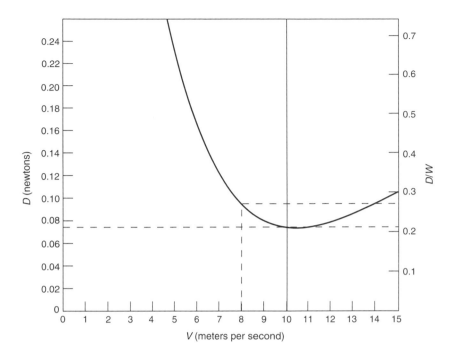

Figure 7 Aerodynamic drag, D, and ratio between drag and weight, D/W, for Tucker's budgerigar. The minimum value of the drag is 0.078 newton (nearly 8 grams), and the lowest value of D/W is 0.22, both at a speed of almost 11 meters per second (25 miles per hour. The speed at which the drag is smallest is higher than the speed at which the power has a minimum (see also figure 6).

budgerigar is 0.078 newton (almost 8 grams) at a speed of 11 meters per second (25 miles per hour), substantially above the speed at which the propulsive power reaches its lowest value. How does this happen? Power is the product of drag and speed (equation 4). Flying slowly is therefore economical if the drag doesn't increase too quickly. The minimum power level obtains at a speed that compromises between a low value of D and a low value of V.

With figures 6 and 7 to choose from, we must now decide which of the two graphs should determine our thinking about the energy consumption of birds and airplanes. In chapter 1 it was cheerfully announced that an optimum cruising speed exists and that all calculations would be based on that optimum. Now comes the question: Should we use the lowest value of P, or the lowest value of D? The answer depends on what we wish to achieve. If we want to achieve the longest flight duration (as would be the case when an airplane is locked in a holding pattern while waiting for a landing slot), we need to minimize energy consumption per unit time. Because power is

Herring gull (*Larus argentatus*): W = 9.4 N, S =0.18 m^2, b = 1.4 m.

energy per second, we can remain airborne longest at the lowest value of P. When time is the decisive factor, figure 6 is appropriate.

If P is the energy required per second, how much energy is consumed per meter? This is, of course, the quantity we want to minimize in order to maximize the distance we travel. The cruising speed we are looking for is the speed at which the energy consumed per meter is as low as possible. Table 2 is helpful here: energy equals force times distance, so force equals energy per unit distance. Forces can be measured in newtons, but joules per meter are just as good. In our case this means that the aerodynamic drag D is identical to the energy consumption per meter traveled. Flying at a speed that minimizes D, we have automatically minimized the energy consumed per unit distance. The cruising speed is the speed at which the smallest value of D is obtained. When distance is the decisive factor, figure 7 is the appropriate graph.

Where does this lead us? The lowest energy consumption per unit distance is achieved at a relatively *high* speed. Birds and planes must fly *fast* to be economical. What a windfall for the automotive lobby if that were true for cars and trucks, too! Imagine that cars were required to travel faster than 70 miles per hour to keep pollution and fuel consumption within limits, or that truckers were fined for going slower than 55! For that matter, just compare Tucker's budgerigar to any land animal of comparable size. The cruising speed of the little bird was 25 miles per hour—well beyond the reach of mice, chipmunks, and squirrels. A rabbit can run for its life if it has to, and can reach speeds close to a budgy's for a few tens of a second, but you can't call that long-distance cruising. No; birds and planes are different: they do it best at speed.

Precisely because airplanes and birds are so different on this score, it is worth the trouble to find out what figures 6 and 7 would look like if they were used to plot the performance of cars rather than birds. The equivalent of figure 6 is straightforward: you can find the necessary data neatly tabulated in any automotive yearbook. You know from experience that you pay dearly to drive faster than your neighbor. Fiat's tiny Panda, with a 20-kilowatt engine and a top speed of 75 miles per hour, would be fast enough for U.S. highways. If you want to drive 100 miles per hour, you need at least 50 kilowatts. A speed of 200 miles per hour requires 300 kilowatts. A Porsche 959 (top speed: 197 miles per hour) has a 331-kilowatt engine. A 911 Turbo makes do with a 221-kilowatt engine, reaching a top speed of a mere 160 miles per hour.

The rate at which the power required increases with increasing speed is illustrated best by another proportional diagram (figure 8). The engine power turns out to be proportional to the third power of the speed. Thus, if you want to drive twice as fast, your engine has to be 8 times as powerful. A Porsche or a Ferrari can go 3 times as fast as a Fiat Panda, but its engine is 27 times as powerful. If you are determined to travel at high speed, you should take up flying. A sport plane with a large Porsche engine can easily reach 250 miles per hour.

Since the power P is equal to the drag D times the speed V, the data in figure 8 can be used to make a graph of the relation between D and V for cars, equivalent to figure 7. But we can do better. It is obvious that large birds and large vehicles consume more food or fuel than small ones. If you want to compare the energy needs of different modes of transportation, you need to account for differences in size. A trucker would advise you to compute fuel consumption not per mile but per ton-mile. And drag is not the most relevant measure for the specific cost of locomotion; a better measure is the drag per unit weight, D/W. This is the quantity we shall focus on, bearing in mind that the payload of a vehicle is often only a fraction of its gross weight.

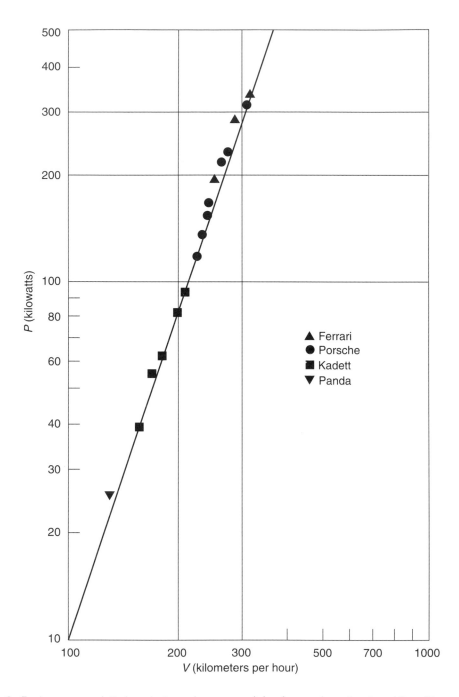

Figure 8 Engine power plotted against maximum speed for four makes of automobiles. The diagonal is steep because power increases as the third power of the speed. To go twice as fast requires an eightfold increase in engine power.

The quantity D/W registers energy consumption per meter traveled for each newton of gross weight. In thinking about the performance characteristics of different modes of transportation, this quantity is so important that it has a separate name: E, the specific energy consumption. In engineering notation:

$$E = \frac{D}{W} = \frac{P}{WV}. \tag{6}$$

The quantity E is "nondimensional": if we stay with the metric system and measure both D and W in newtons, E is a pure number. In the same way, P must be measured in watts and V in meters per second; otherwise one becomes hopelessly tangled in conversion factors.

Since the performance of Tucker's budgerigar was always measured at the same weight (35 grams, or 0.35 newton), figure 7 remains unchanged if we convert the graph from D to E. The only change is the scale on the vertical axis. The scale for D is given on the left side of figure 7; that for E is given on the right. The best value for E that Tucker's budgy could attain was 0.22, at a speed of almost 11 meters per second.

Cars can be dealt with in the same way. Consulting the automotive yearbook once more, and adding 200 kilograms to the empty weight for the driver, one passenger, fuel, and luggage, we can easily convert the data of figure 8 into data for E. The results are presented in figure 9, along with the corresponding numbers for passenger trains (commuter trains, the Amtrak Metroliner, and France's TGV (*train à grande vitesse*). Tucker's budgy and the Boeing 747 are not forgotten, either.

There is plenty of information in figure 9. For one thing, trains would appear to be far more economical than cars. However, when we account for the difference between useful load and gross weight, the economy of trains is not so clear-cut. A two-unit commuter train weighs 100 tons and seats roughly 120 passengers. At 70 kilograms a head, the people in the train weigh about 8.4 tons—only about 8 percent of the gross weight. A Chevrolet Corsica, with an empty weight of 2000 pounds, can carry four people, for a useful load of 600 pounds. This makes the payload equal to 23 percent of the gross weight ($600/(600 + 2000) = 0.23$)—3 times the value for a train. At a speed of 75 miles per hour, a car's specific energy consumption, E, is about 0.08, and that for a train 0.011. Giving the train a markup for its excessive empty weight, we find that the corrected value of E is about

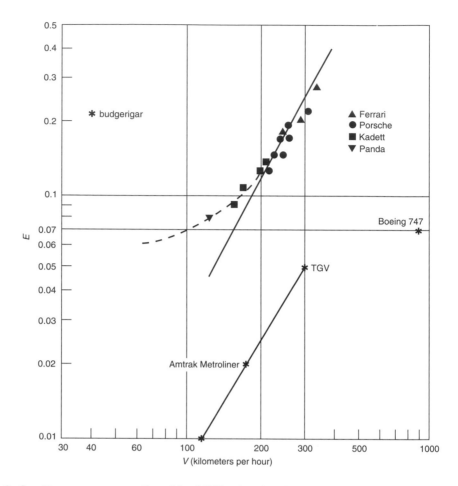

Figure 9 Specific energy consumption, E (= P/WV), plotted against speed.

0.033, roughly half that of a car. These values reflect relative energy consumption per passenger-mile. Here are two easy numbers to remember: a train needs 1.6 megajoules per passenger mile (1 megajoule per passenger-kilometer), and a car twice that. But remember also that in terms of efficiency there is no difference between a fully loaded automobile and a half-empty train! Because the difference between cars and trains is not at all as significant as is often stated, the Netherlands State Railways sell discounted taxicab tickets to their passengers. On the last leg of a journey, a Dutch citizen can now get a $4 ride in a brand-new Mercedes. A marriage between commuter trains and taxis sounds crazy but makes perfect sense.

Back to birds and planes. Tucker's budgerigar can be found in the top left corner of figure 9, which shows that it is rather uneconomical. But budgies are somewhat clumsy flyers. Seagulls achieve a far better value of E, namely 0.1, which compared to cars is really quite

Greylag goose (*Anser anser*): W = 31 N, S = 0.27 m^2, b = 1.63 m.

good. But the most striking feature of figure 9 is that flying becomes less rather than more expensive as speed increases. For instance, the Boeing 747 and other jetliners achieve E = 0.07 at speeds of 560 miles per hour. A straight line drawn in figure 9 from the budgy to the 747 intersects the curve for cars in the vicinity of 200 kilometers per hour. Below that speed wheels perform better than wings, but at higher speeds the wings have it. Flying is the preferred mode of transportation when high speeds are desired. Most vehicles need additional power to achieve higher speeds, but a well-designed airplane can fly fast without using more fuel.

Nutrition and Combustion

So far we have not bothered to make a systematic distinction between fuel consumption and energy consumption, but now the time has come. Different foodstuffs and fuels supply different quantities of energy per kilogram consumed. There are also variations in the efficiencies of different energy-conversion processes. For example, a steam locomotive has a mere 5 percent efficiency, which means that of every 20 joules of energy available in anthracite or fuel oil only 1 joule is delivered as useful work to the driving wheels; the rest literally goes up in smoke. A gasoline engine achieves a conversion efficiency of 25 percent at average speeds; nevertheless, three out of four joules of combustion heat are expelled by the radiator or the exhaust pipe, and only one out of four is available for propulsion. At 30 percent efficiency, diesel engines are a bit more economical.

Coal-fired and gas-fired electricity-generating plants manage 35 percent. The energy-conversion efficiency of human and animal metabolism is typically about 25 percent.

Consider an electricity-generating plant that uses natural gas with a combustion value of 36 megajoules per cubic meter. The purchase price of natural gas is about $5 per 1000 cubic feet, or 18 cents per cubic meter. This works out at half a cent per megajoule. Because two-thirds of the energy is wasted through the chimney or the cooling towers, the selling price of energy must be at least 1.5 cents per megajoule. The user's monthly bill, however, is calculated in kilowatt-hours. A kilowatt-hour equals 3.6 megajoules; thus, in order for the utility company to break even, the selling price of electricity has to be about 6 cents per kilowatt-hour. (Power companies calculate their off-peak rates along these lines. If investment costs can be recovered during peak periods, they are content if only fuel costs are covered at night.)

A megajoule is 0.28 kilowatt-hour. This strange unit comes to life only when we work everyday examples. For instance, reheating some leftover lasagna in a 700-watt microwave oven for 5 minutes uses about 0.2 megajoule of energy, at a cost of about one cent at the off-peak rate. The manual labor of cleaning up one's back yard requires about half a megajoule every hour if one isn't working too strenuously—500,000 joules per 3600 seconds, the equivalent of 140 watts.

The nutritional value of peanut butter is stated on the label of the jar. Peanut butter is good for 180 "calories" per ounce. (These are actually kilocalories, so we are really talking about 180,000 calories per ounce.) With 4.2 joules per calorie and 28.4 grams per ounce, this works out to 2700 kilojoules per 100 grams, or 27 megajoules per kilogram. Thus, a megajoule of peanut butter weighs about 37 grams, enough for three solid sandwiches. The price of this megajoule is approximately 10 cents, substantially more than the price of a megajoule of electricity. Electric trains use 1.6 megajoules per passenger-mile. (Just for fun, and ignoring the resulting mess, you can work out how much it would cost to run the train on peanut butter.)

The carton of milk you buy at the supermarket carries nutritional information, too. Milk supplies 600 kilocalories per quart (280 kilojoules of energy per 100 grams, or 10 megajoules per U.S. gallon). If milk costs $2 a gallon, a megajoule will set you back 20 cents. When you find out that the price of natural gas is something like half a cent per megajoule, you might almost be tempted to drink liquefied methane instead of milk. Steak is an outrage from this perspective. Its nutritional value is 4 megajoules per kilogram, roughly 2 megajoules per pound. At a price of $10 a pound for prime beef, we are talking about $5 per megajoule. Even the cheapest cuts of beef, at $3 a pound, cost as much as $1.50 per megajoule. Delicious, but outrageously expensive.

To enable you to do more calculations of this type, table 3 presents the nutritional values of several common digestibles and the heats of combustion of several fuels. In dollars per megajoule, natural gas and gasoline stand out as best buys. At 30 cents a liter ($1.15 per gallon), gasoline in the United States costs about one cent per megajoule. (In Europe, gasoline costs about $4 per gallon, two-thirds of which goes to taxes.)

Among digestibles, vegetable oil is cheapest by far. Salad oil is barely digestible, and consumed in quantity it results in a bad heartburn. Peanut butter, though, is an excellent alternative for those who insist on getting a megajoule for their buck.

Table 3 The heat of combustion or metabolic equivalent for various foodstuffs and fuels. The prices are based on a "snapshot" in 1994; large fluctuations may, of course, occur over time.

	MJ/kg[a]	$/kg	$/MJ	Comments
Prime beef	4.0	20	5	
Beef	4.0	8	2	
Whole milk	2.8	0.90	0.32	600 cal/quart
Honey	14	4	0.29	
Sugar	15	1	0.07	100 cal/ounce
Cheese	15	6	0.40	
Bacon	29	4	0.14	
Corn flakes	15	3.50	0.23	100 cal/ounce
Peanut butter	27	4	0.15	180 cal/ounce
Butter	32	4.50	0.14	
Vegetable oil	36	2	0.06	240 cal/ounce
Kerosene	42	0.40	0.010	0.82 kg/liter
Diesel oil	42	0.40	0.010	0.85 kg/liter
Gasoline	42	0.40	0.010	0.75 kg/liter
Natural gas	45	0.24	0.005	0.8 kg/m^3

a. megajoules per kilogram

Snow goose (*Chen hyperborea*).

How much food do birds need when they migrate south in the fall? A mute swan, the largest of the European swans, flies at a cruising speed of some 24 meters per second (54 miles per hour). It weighs 10 kilograms (22 pounds), 2 kilograms of which constitute its flight muscles. In cruising flight, the power output of these flight muscles is about 100 watts per kilogram. During long-distance travel, therefore, the flight muscles of a swan supply approximately 200 watts of mechanical power. Because the conversion efficiency of nutrient energy is only about 25 percent, the swan needs to consume 4 times as much nutrient energy: 800 watts, or 800 joules per second. At a speed of 24 meters per second, this corresponds to an energy consumption of 33 joules per meter, or 53 kilojoules per mile. This energy is supplied by the spare fat on the bird's chest.

During a long flight, the pectoral muscles of birds metabolize fats directly. Human muscles, in contrast, burn sugars. In the human body, the liver must convert fats into sugars before the stored energy is of any use to the muscles. At the high metabolic rates typical of birds—remember, flying is plain hard work in most cases—this is not an attractive option. On top of that, the nutritional value of fat is twice that of sugar (table 3). Thus, it is much better to carry fat for fuel rather than sugar.

As always, there are exceptions. Hummingbirds run on honey and sugar water, and some marathon runners manage to convert from sugar to fat metabolism in the course of a race (a process that causes severely painful physiological changes). Most insects run on sugars

White pelican (*Pelecanus erythrorhynchos*): W = 60 N, S = 1 m^2, b = 2.80 m.

(just think of the honeybee), and so do chickens and game birds (which are not capable of long-distance flight). Migrating butterflies, however, store fat in their abdomens.

The nutritional value of bird fat is roughly equal to that of butter: about 30 megajoules per kilogram, or 30 joules per gram. Since a swan requires 33 joules per meter, or 53 kilojoules per mile, it consumes 1.1 gram of fat per kilometer (a little under 2 grams per mile). After 12 hours of cruising at 54 miles per hour, the bird has clocked 650 miles and has lost more than a kilogram of body weight (1150 grams to be precise, and not counting the energy needed for its other body functions). Obviously, a light snack will not fill its stomach at the end of such a working day. The same is true for homing pigeons at the end of a long-distance race: they are not only dead-tired, but famished as well. From this perspective, the importance of bird sanctuaries is easy to understand: a migrating bird must eat voraciously before continuing its journey.

Gannet (*Sula bassana*): W = 27 N, S = 0.25 m^2, b = 1.85 m.

The KLM flight that leaves London-Heathrow for Amsterdam at 6 P.M. each day is like an airborne commuter train. The Boeing 737 is filled with regular customers: business people returning home from a day's work, familiar voices chattering to pass the time and teasing the long-suffering flight attendants. A few years ago, as I was dozing in my seat after a tedious meeting at the European Center for Medium-Range Weather Forecasts, the captain's voice on the public address system woke me up: "Ladies and gentlemen, this is the captain speaking. As you know, this trip usually takes about 45 minutes, but today is different. We have a 145-knot tailwind, which corresponds to 170 miles per hour. I have never before encountered such fierce winds. To compensate a little for the turbulence on this trip, we will arrive in Amsterdam 10 minutes early." (A knot is a nautical mile per hour, and a nautical mile equals 1.15 statute miles.) The 737 rode the center of the westerly jet stream, and the pilot played it to full advantage. The airspeed was 500 miles per hour, but because of the tailwind the groundspeed was 670 miles an hour.

When flying you must always take the wind into account, and you can't afford to be casual about it. Obviously it is convenient and economical when a tailwind helps you along, but you should be on guard when coping with headwinds. They set you back, and that cuts into your range. During World War II, American bombers on their way to Japan from Saipan and Tinian (islands just north of Guam) sometimes had to return prematurely because the headwinds they encountered were stronger than had been forecast. (There is a story of one B-29 bomber squadron actually flying backwards on a particularly breezy day: fully opened throttles and 200-plus miles per hour were not sufficient to make headway against the storm.) Having consumed more than half the fuel in their tanks, they had no choice but to return. To lighten their planes and thus reduce fuel consumption, the crews dumped their bombs in the ocean. Once they had turned around, the headwinds became tailwinds, and so they got home in a hurry.

Wind influences airlines' timetables, too. The flight from Amsterdam to Los Angeles takes about an hour longer than the return trip. Flying at 30,000 feet and up, jetliners encounter stiff westerlies most of the time. Those mid-latitude winds are caused by the interaction of the earth's rotation and the temperature contrast between the equator and the poles. At cruising altitude, the average wind speed is about 30 miles per hour. Hence, during the 10-hour journey between Amsterdam and Los Angeles you lose 300 miles. Since long-distance jetliners travel about 550 miles per hour, this adds a little more than 30 minutes to the flight. Flying in the other direction, the wind works to your advantage. This explains the one-hour difference in the timetable.

The cooperation between meteorology and aviation benefits both parties. Airliners participate in the observation and measurement programs of worldwide meteorology, and the weather computers predict where the winds will be strongest. Every day, the most economical routes across the oceans are selected in international conference calls of air traffic controllers. Westbound traffic often makes appreciable detours in order to avoid the strongest of the forecast headwinds. If the turbulence is not too severe, eastbound traffic is directed into the heart of the westerly jetstream. These adjustments reduce both travel time and fuel consumption. In the crowded skies above Europe and the United States, however, airline traffic is so congested that everyone must stick to the appointed airways. Only coast-to-coast nonstop flights are assigned the routes with the best winds.

The slower you fly, the more the wind will affect you. The large propeller-driven airliners of the 1950s, the Douglas DC-7 and the Lockheed Constellation, cruised at about 300 miles per hour and had to stop for fuel at Gander, Newfoundland, if they ran into unexpected headwinds over the Atlantic. These days, with cruising speeds twice as high, stopping at Gander is a thing of the past.

What is true for airplanes is also true for birds. Because birds are much smaller than planes, they fly more slowly (see chapter 1), and this makes them vulnerable to adverse weather. The average wind speed on earth is between 5 and 10 meters per second (11 and 22 miles per hour). If they want to return home as a storm approaches, birds must be able to fly about 10 meters per second, and their wing loading must range between 10 and 100 newtons per square meter.

White stork (*Ciconia alba*): W = 34 N, S = 0.5 m^2, b = 2 m.

(See figure 2.) Most birds weigh between 10 grams and 10 kilograms, which puts their wing loading in the desired range. The vertical line in the center of figure 2 was drawn for a good reason: birds that fly faster have some speed to spare when the wind increases; the rest must watch out.

Still, one can easily imagine circumstances in which the wind blows so hard that all birds must seek shelter. This happens sooner for small birds than for large ones (see table 4). Storm petrels were so named because they are the first ocean birds to seek refuge ashore as a storm moves into the area. Their arrival above land is an early warning signal. Small birds have low wing loadings and hence low airspeeds. As a consequence, they must seek shelter sooner than their larger relatives. The very smallest birds, goldcrests and kinglets, cannot survive the open plains or the ocean shores. Their habitat is the forest, where trees and shrubs protect them from high winds.

Insects are so small that their lives are dominated by the wind, and the very smallest must wait for the wind to die down at sunset before taking to the air. Before gnats start their dance in your back yard early in the evening, they have done some solid thinking. The insects in figure 2 can be divided into two categories: those with high and those with low wing loadings. Beetles, flies, bees, and wasps belong to the first category, butterflies and dragonflies to the second. Honeybees can return to their hive after harvesting nectar and pollen in a faraway rapeseed or alfalfa field. Butterflies, however, must accept that they may be blown away by the wind, and dragonflies cannot

Table 4 Wind and airspeed. Wind speed is given in meters per second and according to the Beaufort scale. The cruising speeds of various insects, birds, and airplanes are given for comparison.

m/sec	Beaufort number		Airspeed
— 0.6 — — — 1	1	Light air	Butterflies
— 2 — 3	2	Light breeze	Gnats, midges, damselflies
— 4 — 5	3	Gentle breeze	Human-powered aircraft, flies, dragonflies
— 6 — — 8	4	Moderate breeze	Bees, wasps, beetles, hummingbirds, swallows
— 10	5	Fresh breeze	Sparrows, thrushes, finches, owls, buzzards
	6	Strong breeze	Blackbirds, crows
	7	Near gale	Gulls, falcons
— 20	8	Gale	Ducks, geese
	9	Strong gale	Swans, coots
	10	Storm	Sailplanes
— 30	11	Violent storm	Light aircraft
	12	Hurricane	

venture far from their hideaways in windy weather. But accidents do happen occasionally—for example, grasshoppers from the Sahara sometimes turn up in England.

The difference between a maritime climate and a continental one is considerable. Ocean birds live in a windy environment, which explains why most of them are fairly large. Large birds have higher wing loadings and airspeeds than small ones, and this is a major advantage at sea. Because ocean birds must cover great distances, it is essential that their energy consumption per mile be low. Their long, slender wings achieve just this. The narrow wings of seagulls, terns,

Swallowtail (*Papilio machaon*): W = 0.006 N, S = 0.003 m^2, b = 0.08 m.

skuas, and albatrosses are quite different from the broad wings of vultures, condors, and eagles. Those large birds of prey do not travel long distances; they soar in circles, taking advantage of the ascending motion in hot bubbles of air ("thermals," in glider pilot jargon). In thermal soaring the energy consumption per unit distance does not matter. The wings of soaring birds of prey minimize the energy consumption per unit time (that is, the energy that must be extracted from the air). This goal is achieved by flying slowly; hence the low wing loadings. The enormous wings of the golden eagle make perfect sense.

The Art of Soaring

Flying is an arduous way of life. This is why several species of birds have discovered how to stay airborne without flapping their wings. The trick is to find upward air movements of sufficient strength. Under normal circumstances a bird glides down when it doesn't flap its wings; then, as it loses altitude, gravity supplies the energy needed to maintain airspeed. But when the rate at which the wind lifts a bird is equal to its rate of descent, the bird can stay up indefinitely. And if the upward air motion is stronger than the bird's sinking speed, it can gain altitude if it wishes. Staying aloft without having to work for it: that is the art of soaring.

There are several ways to soar. One is practiced by herring gulls as they follow a ferry or a cruise ship. They fly on the windward side of the ship, where the wind escapes upward as it strikes the superstructure. The gulls need only adjust their wing area. When the wind increases, they fold their wings a little. If they lose too much altitude, they spread their wings again. If they start going too fast, they simply extend their feet a bit more, and that slows them down. Webbed feet

Heron (*Ardea cinerea*): W = 14 N, S = 0.36 m^2, b = 1.70 m.

are perfect air brakes. (Every glider pilot knows how important it is to increase drag briefly when flying too fast or too high.) The flight of seagulls alongside a ferry is called "slope soaring." It can also be practiced along chains of dunes and mountain ridges. But there must be sufficient wind, because the upward velocity along the slope is proportional to the wind speed. Furthermore, the wind should blow perpendicular to the ridge or it will not be diverted upward (figure 10).

In favorable circumstances, gulls and terns can soar back and forth for very long periods without ever flapping their wings. They are getting a free ride, and what a ride it is! Kestrels and harriers (sparrow hawks and marsh hawks to some) do the same inland, along the slopes of hills and levees. Hang gliders and sailplanes also take advantage of slope winds. Since hang gliders have a sinking speed of more than 2 meters per second, they need a stiff breeze before they can venture flying along the shore.

Slope soaring is perfect for covering large distances. Glider pilots achieve their distance records on days when high winds are blowing across long mountain ridges. The narrow northeast-southwest folds of the Appalachian Mountains stretch from Elmira, New York, to

Figure 10 Slope soaring alongside a ferry and along a dune ridge.

Chattanooga, Tennessee. When a storm hits that 600-mile stretch, stirring up blustery northwesterly winds behind its cold front, complete with deep-blue skies and towering cumulus clouds, glider pilots in Elmira ("soaring capital of the world") just can't wait to get airborne.

The second method of soaring capitalizes on thermals (ascending pockets of hot air). This is not without its drawbacks. For one thing, thermals do not occur everywhere, and you are not going to get anywhere unless you can first find a thermal in which you can gain altitude. You do that by circling in the hot, rising column of air. Once you have climbed the thermal for a while, you can start your journey by gliding in the general direction of your destination, hoping to find the next thermal before losing too much altitude. The flight toward

Kestrel (*Falco tinnunculus*): $W = 1.8$ N, $S = 0.06$ m^2, $b = 0.74$ m.

Flamingo (*Phoenicopterus ruber*).

your goal will be punctuated by such episodes in the winding staircases of rising air.

This kind of soaring is possible only during the day, because thermal motion occurs only after the sun has started heating up the earth's surface. You can observe this by watching buzzards (buteos, to some) and other soaring birds of prey. As the morning progresses, these birds test the strength of the convection currents. They take wing, searching for ascending air. If they fail to maintain altitude, they return to their tree or rock ledge and wait for the earth to warm up a bit more. The flight muscles of soaring birds of prey are not very

Osprey (*Pandion haliaetus*): W = 15 N, S = 0.3 m^2, b = 1.60 m.

powerful and cannot sustain flapping flight for more than a few minutes. They have no choice but to wait.

As hot air rises, it cools—about 1° Celsius per 100 meters. At a certain altitude, the water vapor in the ascending air begins to condense; the cumulus clouds created that way are a sure sign of upward motion. For this reason, glider pilots join the gulls and hawks that are circling below these clouds when they want to gain altitude. The rates of descent of soaring birds and gliders are comparable (around 1 meter per second), but the airspeed of birds is lower. Birds can keep up with their fiberglass companions only by flying in tighter circles. If a bird and a glider pilot both are in a playful mood, or have a researcher's attitude (which comes to the same thing), they may start flying competitively to see who dares to turn the tighter circles, who can fly slower without losing control or stalling (literally dropping out of the race in that case), who has the smarter tactics for discerning the next thermal and reaching it with minimum altitude loss, and so on. Buzzards appear to take these games very seriously. When the soaring habits of griffon vultures in Central Africa were investigated by a biologist in a glider with auxiliary engine nature, science, and technology were united in the skies over the savannah on a hot summer day.

The Great Migration

Birds cover enormous distances on their annual migrations. They must make professional weather forecasts or risk running into serious trouble. Since their cruising speeds are relatively low, they will consume too much fuel if they run into headwinds. When the wind direction shifts, they risk being blown off course, with fatal consequences if they should end up over the open ocean. Careful preparations are necessary before they start on their journey. This is especially true for small birds, such as the North American passerines that cross the Gulf of Mexico on their way south—chimney swifts, bank and cliff swallows, purple martins, blackpoll warblers, redstarts, and the like. Their performance should not be underestimated, even if the 500-mile trip across the Gulf is not quite as far as the 1000-mile crossing of the Sahara Desert. But the real heroes of the North American migration business are the monarch butterfly and the ruby-throated hummingbird. The monarch is known to consume fat on its way across the Gulf, but how about the ruby-throat? Does it burn fat or sugar on long-distance flights?

Black-capped chickadee (*Parus atricapillus*): W = 0.12 N, S = 0.0076 m², b = 0.21 m.

The little European passerines that migrate to Africa for the winter, crossing both the Mediterranean and the Sahara on their way, are courageous birds, too. They include various kinds of warblers, wagtails, pipits, chiffchaffs, flycatchers, whinchats, redstarts, and nightingales. With a cruising speed of only 7 meters per second (16 miles per hour), they must have adequate tailwinds before they can start their crossing. Little birds consume relatively large amounts of energy. For passerines, just as for Tucker's budgy, the specific energy consumption ($E = D/W$) is approximately 0.25. At cruising speed, therefore, their aerodynamic drag is about one-fourth of their weight, which is not particularly economical.

Because all these birds cross the Sahara without fuel stops, their cruising range must be at least 1000 miles. They manage this by storing so much fat on their chests that they can barely fly. A 20-gram wagtail, with a normal fuel reserve of 5 grams, starts its journey across the Sahara with an additional 10 grams of fat. Its takeoff weight is 30 grams (a little more than an ounce)—twice its zero-fuel weight. Half of the takeoff weight is fuel, much the same as for a long-distance airliner. Because these little birds must fatten themselves in preparation for their flight across the Mediterranean, and have trouble taking off with so much excess weight, they are easy prey for Italian bird catchers. (Unfortunately, a roasted nightingale tastes much better when its meat is wrapped in fatty tissue.)

Let's take the average weight of a wagtail on a long-distance flight to be 24 grams, and assume that its drag is a fourth of its weight. The average drag then is 6 grams, or 0.06 newton. As was explained in chapter 2, a newton is a joule per meter. Thus, if we can compute how many joules of mechanical energy are supplied by metabolizing 15 grams of bird fat, we can calculate the wagtail's maximum range. Bird fat supplies 32 kilojoules per gram (chapter 2); hence, 15 grams supply 480 kilojoules. However, since the bird's metabolic efficiency is only about 25 percent, the net supply of energy is no more than 120 kilojoules. This is used up at a rate of 0.06 joule per meter, or 60 kilojoules per kilometer. After 2000 kilometers (1250 miles), all the fat is gone.

With fuel reserves for a mere 250 miles on a 1000-mile trip, there is not much to spare. Not even airliners cut it that close! Birds cannot afford any miscalculations in their weather forecast. They are wise enough to wait for the wind to blow in the right direction. Circling over their feeding grounds each morning, they check the meteorological conditions. The starting signal for the great journey is given only when conditions are favorable. Even with a fair tailwind, the flight across the Sahara takes at least two days and a night; if the weather suddenly deteriorates during the trip, massive mortality may result. Similar risks are taken in crossing the Gulf of Mexico. Few passerines survive when they are blown far out into the Atlantic by unexpected westerly gales. Through the centuries sailors have told stories of songbirds escaping their fate by dropping on the deck of a ship, famished and exhausted, more dead than alive, but recuperating quickly on bread crumbs, scraps of bacon, and a sailor's tender loving care.

Migrating ocean birds cover distances much greater than the 500 miles across the Gulf of Mexico or the 1000 miles across the Sahara. Several species of plovers, godwits, and sandpipers make nonstop trips from Cape Cod to Trinidad, a distance of 2500 miles and pretty close to their maximum range. Since this kind of flying requires much more advanced aerodynamics, the bodies of ocean birds are streamlined and their wings slender. As a result, the ratio between drag and lift decreases substantially: for ocean birds it is about 0.1, instead of the 0.25 typical of songbirds. Wilson's phalarope, a little shore bird that migrates more than 5000 miles along the Pacific coast of the Americas, prepares for its annual journey by filling its belly with brine shrimp in Lake Mono, California. Like migrating passerines, it fattens itself until it can barely fly. Because the ratio $E = D/W$ remains the same, its aerodynamic drag increases by 50 percent

Ruddy turnstone (*Arenaria interpres*).

when its weight grows to 50 percent above normal. Carrying that much excess weight, yet using the same wings, the phalarope must fly more than 20 percent faster ($\sqrt{1.5} = 1.225$). The power required equals drag times speed ($P = DV$; see chapter 2); it increases to almost twice the normal value ($1.5 \times 1.225 = 1.84$). If a phalarope continued stuffing itself, it wouldn't be able to take off at all. Once airborne, however, with such an economical value of E, it flies more than 3000 miles nonstop. Even so, it cannot make the journey from California to Chile without stopping for fuel along the way. From time to time it has to forage for shrimp and other seafood along the beach.

Each species has its own strategy. The sandpipers that breed along the coast of Greenland and pass the beaches of Northwestern Europe each August, en route to destinations further south, appear to have learned that westerly winds are generally stronger at high altitudes. On their 1200-mile journey to Scotland, they have been observed flying as high as 7 kilometers (23,000 feet)—almost above the weather. For the same reason they skim the waves on their journey back to Greenland, riding on the easterly surface winds blowing north of mid-latitude storms.

Taking Off and Landing

Whenever they can, birds and airplanes take off and land into the wind. They need speed in order to become airborne. It is the speed with respect to the air that matters, not the speed with respect to the ground. Pilots speak of airspeed versus groundspeed. When the wind blows, groundspeed and airspeed are not the same. Franklin's gull and the similar European black-headed gull have an airspeed of nearly 10 meters per second. That is 22 miles per hour, or force 5 on the Beaufort scale. With a headwind of 10 meters per second, a gull makes no headway at all. This is a nuisance if one has to get somewhere, but it becomes an advantage during takeoff and landing. A gull perched on a warehouse roof or a harbor bollard in a stiff breeze has only to spread its wings to obtain the lift required. A little hop into the air and a few casual wingbeats and away it flies, with no effort at all. Taking off in calm weather is not so easy. The gull can either dive off its perch or take off vertically with rapid beating of its wings. The second option is hard work, requiring 4 times as much power as ordinary flight. This is why most birds prefer to take off from a tree, a telephone pole, a gutter, or some other elevated object. Starting with a brief dive, the bird gains the necessary airspeed by letting gravity do the work.

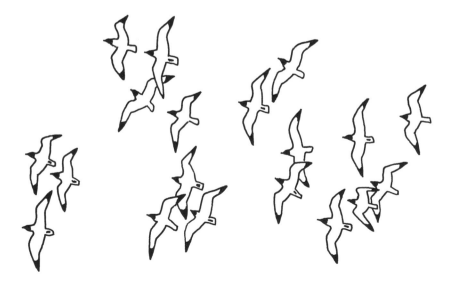

Franklin's gull (*Larus pipixcan*): W = 2.5 N, S = 0.08 m², b = 0.95 m.

Rock dove (domestic pigeon, *Columba livia*): W = 2.8 N, S = 0.07 m², b = 0.78 m.

Landing works in much the same way. If circumstances allow, a bird lands into the wind, because that minimizes groundspeed. This is why a bird can land on a fence quite casually, as though it doesn't require great precision and exacting coordination. A pigeon lands on a roof ledge or a windowsill by deliberately approaching the landing spot from below and sailing gracefully upward in the last few meters, losing speed on the way up until its flight speed drops to zero at the chosen spot.

If a bird has to cope with a tailwind when taking off, it is in trouble. Because its groundspeed is now higher than its airspeed, it must run like crazy before its airspeed is high enough for takeoff. When the wind comes from behind, a bird must make an extremely long takeoff or landing run.

Airliners also have to worry about the wind, though not as much as birds, because their speeds at takeoff and touchdown—about 190 and 130 miles per hour, respectively—are much faster than typical windspeeds. Nevertheless, no airplane can afford to take off or to land with the wind at its back.

The runways of major airports are usually about 2 miles long. Is that long enough for the takeoff run of a wide-body jet? An airplane can lift itself off the ground only after achieving sufficient airspeed. It must be accelerated before it can fly, and to do the necessary calculations we need to know the plane's acceleration when all its engines are running at takeoff power. The jet engines of a modern airliner deliver a total thrust equal to roughly one-fourth of the takeoff weight. Not all of this thrust can be used for acceleration, however; we have to make an allowance for the average aerodynamic resistance during the takeoff run. Therefore, we estimate the net thrust to be 20 percent of the takeoff weight. With T standing for thrust and D for drag, we have

$$T - D = 0.2W. \tag{7}$$

Because force times distance equals work, the product of the net thrust and the length R of the takeoff run is the work performed by the engines. This work is converted without any loss into the kinetic energy of the airplane. You may remember from your high school science class that the energy of motion, or kinetic energy, is $\frac{1}{2}mV^2$, where m is the moving object's mass and V is its speed. The last thing we need here is the relation between mass and weight. The weight W is the force exerted on an object of mass m by the pull of gravity. If we call the acceleration of gravity g, we can write

$$W = mg. \tag{8}$$

With the aid of equation 8, we can write the kinetic energy $K = \frac{1}{2}mV^2$ as

$$K = \tfrac{1}{2}mV^2 = \tfrac{1}{2}(W/g)V^2. \tag{9}$$

The energy supplied by the engines equals the net thrust times the length R of the takeoff run. Using equations 7 and 9, we obtain

$$\tfrac{1}{2}(W/g)V^2 = 0.2WR. \tag{10}$$

This can be simplified. When we divide both sides of equation 10 by W and multiply both sides by $2g$, we find

$$V^2 = 0.4gR. \tag{11}$$

Computing the length of the takeoff run now becomes a cinch. With $g = 10$ meters per second squared and $V = 84$ meters per second (190 miles per hour), we obtain $R = 1764$ meters (almost 5800 feet; a little more than a mile). However, in order to provide an adequate

Sandhill crane (*Grus canadensis*).

margin of safety, a runway must be roughly twice as long as the takeoff run. Should one of the engines fail during takeoff, an airplane should still be able to come to a complete stop at the far end of the runway. Therefore, an airplane requiring a 1-mile takeoff run needs a 2-mile runway. This, incidentally, is a good example of the design philosophy used in aviation safety calculations. Fair margins, neither too large nor too small, are incorporated to allow for adversities that, fortunately, very rarely arise.

When we start adding realistic details, calculations such as these become rather involved. The maximum speed at which a takeoff run can be safely aborted depends on the gross weight of the airplane and several other factors. The copilot consults the aircraft manual to find the precise number, and when the "decision speed" is reached he or she lets the captain know. Beyond this point it is impossible to brake to a full stop. The plane is now committed to taking off, even if one of the engines fails. A few seconds after reaching decision speed, the pilot pulls the nose of the airplane up, and a few seconds after that the airplane leaves the ground. (The risk of engine failure is one of the reasons why most intercontinental airliners have three or four engines. A 747 with three of its four engines working properly can still take off safely, though it may not be able to climb very fast. It is a tribute to the tremendous reliability of modern jet engines that several kinds of two-engined airliners are allowed to cross the oceans.)

For more heavily loaded airplanes, these constraints become narrower. A 747-400 with a takeoff weight of 380 tons needs to accelerate to 210 miles per hour before it can become airborne—20 miles per hour more than the figure we used a moment ago. With $V = 93$ meters per second (210 miles per hour) instead of 84 meters per second (190 miles per hour), we compute $R = 2160$ meters (almost 7100 feet). At that point a 10,000-foot runway has only 2900 feet left. No wonder that the decision speed is much lower than the takeoff speed in this case: only 180 miles per hour. At this point, some 5000 feet of runway have disappeared under the wheels.

It is easy to compute how wind speed affects a takeoff run. If there is a 30-mile headwind, the airspeed of 210 miles per hour needed by a fully loaded 747 for liftoff is reached at a groundspeed of 180 miles per hour. The takeoff run is then reduced from 7100 to 5200 feet. While you are checking this, using equation 11, please take a moment to consider the opposite situation. With a 30-mph tailwind during takeoff, the groundspeed would have to be 210 + 30 = 240 mph before the airspeed would be high enough for liftoff, requiring a ground run of 9300 feet. This would leave precious little runway to spare. (Relax; no pilot would ever try this.)

It is also advantageous to land into the wind, of course. In a 30-mph headwind a Fokker F100, with a landing speed of 120 mph, has a groundspeed of only 90 mph when it touches down, shortening the landing run considerably. For this reason, air traffic in the vicinity of an airport is always arranged in such a way that all aircraft take off and land into the wind. Diverting traffic to the runway causing the least noise pollution is only possible when there is little wind. As airliners preparing to land at London Heathrow skim their rooftops, the citizens of the London suburb of Hounslow are forcefully reminded of the prevailing westerly winds.

White stork (*Ciconia alba*): W = 34 N, S = 0.5 m^2, b = 2 m.

Approach Procedures

Rigid procedures must be observed in order to achieve orderly air traffic around airports. The usual preparation for the landing sequence begins with a descent to 5000 feet. The pilot then receives radar vectors to a point about 12 miles "downwind" (that is, parallel to the intended runway but in the opposite direction). Then the plane makes a 180° "procedure turn" to align itself with the runway. After this last turn, flaps and landing gear are extended, and airspeed is reduced to about 150 miles per hour, or 2.5 miles per minute. Since there is a distance of about 12 miles to cover before touchdown, this segment of the flight, which is called the "final approach," takes about 5 minutes. Most passengers wouldn't mind having this part sped up a little, but for pilots the last few minutes before touchdown are very busy. A hurried approach, with a steep turn just before touchdown, would cause great stress in the cockpit. Stunts like that are best left to fighter pilots and barnstormers.

It is not only pilots who need to practice approach procedures until they become routine. Birds must do the same thing. In my college years I used to go to summer camp on one of the North Sea islands

Cape pigeon (*Daption capensis*): W = 4.3 N, S = 0.077 m^2, b = 0.88 m.

off the Dutch coast. There was plenty of time for birdwatching, and what I saw one afternoon may have been enhanced somewhat by my imagination. A mature herring gull was teaching his fledgling son step by step all that is involved in landing. First, choose your landing site and watch the waves for the wind direction. Next, monitor the crosswinds that drive you off-course and fly downwind for a while before making a turn into the wind. Now real skills are needed. Stop flapping your wings, start your descent at a speed that allows you to cope with wind gusts, monitor your descent with reference to your landing site, extend your legs a little if you are not descending fast enough, reduce your speed during the last few seconds of the approach by leaning back and keeping your nose high, decelerate all the way by spreading your wings and tail fully, lower your landing gear, lift your wings above your shoulders, touch down, and fold your wings.

The juvenile gull, easy to spot because of its mottled gray and brown feathers, did its best to follow father's example, but with mixed success. On that afternoon it already knew the difference between downwind and final approach; however, it couldn't get the rest right, and it made a real mess of it after the final turn. Sometimes its approach was too high, sometimes too low. When its speed was too high, the youngster tried to correct by leaning back rather than extending its legs. What happened then? Lift increased, and the bird would soar upward until it realized that it was well above its glide path. A steep dive followed: wrong again. Diving generates excessive speed. If you try to fix that by pulling up again, you will find yourself too high for a second time. All very reminiscent of a student pilot early in flight training.

This rather undignified performance often culminated in pure embarrassment. Despite its efforts, the young gull did not manage to make a single smooth touchdown. Sometimes it would land too fast and would trip over its feet, performing an accidental somersault. Then it would try to imitate its father's gentle "flare-out." (By leaning back and fully extending its wings and legs just before touching down, a bird loses speed without gaining altitude. At the moment its lift and its speed drop to zero, its feet should be only an inch or so above the beach. Now the bird simply falls, but since it has performed this trick so well it touches the sand with a barely perceptible impact.) But the young gull continued to overreact by flaring out so abruptly that it would lose lift and speed several feet above the beach, then drop like a brick.

A large and cumbersome bird can't maneuver as nimbly as its smaller cousins. This makes it even more important for such a bird to follow correct flight procedures. A herring gull at Fisherman's Wharf in San Francisco has to be wary of the antics of the wind between the piers and the warehouses, but a few rapid wingbeats and a steeply banked turn will get it out of trouble quickly if something unexpected happens. The brown pelicans that also live here must pay much more attention to the wind. A pelican is a large bird, with a weight of 3 kilograms (7 pounds), a wingspan of 2.20 meters (7 feet), and a wing area of approximately 0.5 square meter (5 square feet). Its wing loading is around 60 newtons per square meter, and its cruising speed is about 12 meters per second (27 miles per hour). Because its wings are enormous, a pelican flies no faster than a herring gull or a homing pigeon, though its weight is comparable to that of the common loon, which must cruise at more than 20 meters per second (45 miles per hour) in order to stay in the air.

Having lunch at Fisherman's Wharf several years ago, I watched a mature pelican that wanted to join a small group of other pelicans waiting for the return of the fishing boats that had gone out to sea that morning. The bird had to make its final approach less than a foot above the mastheads of a dozen fishing boats moored alongside a large warehouse. That necessitated a steep descent in the last 100 feet of the flight. Diving down was out of the question, because that would have made the pelican gain speed just when it needed to lose both speed and altitude.

On the first try, everything went wrong. Thirty feet before touchdown the pelican was suddenly blown off course by a vicious crosswind gust—the same kind of mishap that pilots of small planes worry about. The pelican had to shift gears instantly, summoning all the emergency power of its massive wings to perform a steeply

Laughing gull (*Larus atricilla*): W = 3.3 N, S = 0.1 m^2, b = 1 m.

banked climbing turn. It avoided a collision with the landing pier by only inches, shifted back to maximum continuous power, made a procedure turn to the right, and flew 300 feet on the downwind leg of the approach pattern in preparation for the next attempt. Then it made a 180° procedure turn to the right to begin final approach, and skimmed the mastheads of the fishing boats in its steep descent toward the pier. The pelican flew as slow and low as it dared. Fortunately there were no further gusts. The bird extended its feet, reduced its speed to the bare minimum, and flared out. It made a beautiful landing, giving no hint of having required all the skill a pelican can muster. It didn't even have to move its feet; it just turned around and joined its friends.

Many years ago, I was involved in a near miss at Rotterdam Airport. Having recently obtained my private pilot's license, I was practicing takeoffs and landings in a single-engine Saab Safir with the reporting marks PH-UEG ("Echo Golf" in radio communications). In pilot's jargon, this kind of practice is called "touch and go."

I was flying in the traffic pattern, on course and on speed, half a mile downwind of the runway threshold, with everything under control. Suddenly, in my earphones, I heard a sharp command: "Echo Golf, turn left *now*." The "now" meant that the order was to be executed immediately. As I rolled the Safir hard left to begin the required turn, I responded "turning," confirming that I was following the order without delay.

I hadn't noticed it yet, but the traffic controller knew that 15 miles away an airliner had just started its approach. There were still a few minutes to spare, and the man in the control tower evidently thought that there was sufficient room to slip the student pilot on touch-and-

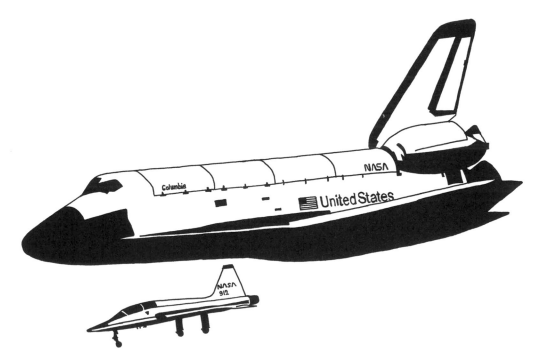

Space Shuttle ($W = 1 \times 10^6$ N, S = 250 m^2, b = 24 m) and Northrop T-38 (W = 1.15×10^5 N, S = 17.3 m^2, b = 8.13 m).

go exercises in front of the distant airliner. Even before I arrived in front of the runway the controller cleared my way: "Echo Golf, cleared to land runway two four." Busy with landing gear, wing flaps, airspeed, carburetor heating, and lots of other details, I continued on "short final," pulling back on the steering column in order to lose speed, extending the wing flaps fully, and closing the throttle all the way.

When I was a few hundred feet from the threshold of the runway, with only seconds to go before touchdown, another airliner suddenly started to taxi onto the runway in preparation for takeoff. The pilot had neither seen nor heard me, and he hadn't waited for clearance from traffic control. (One must always requested and obtain explicit permission before entering an active runway.) Too busy preparing for touchdown, I had not seen him either.

Realizing the danger, I rammed the throttle lever full forward. By doing that I could have killed myself. Airplane piston engines tend to starve from lack of fuel when the throttle is suddenly opened; their carburetors don't have acceleration pumps. The engine hiccupped, and the propeller seemed to stop, but then, thank God, the engine picked up and started to roar. At full power, I just managed to avoid the tail of the airplane that had committed the traffic violation. White around the gills, I flew the traffic circuit once more. Although my touchdown wasn't particularly smooth, I was happy to be on the ground again. Still trembling, I rode back home on my ancient moped. All was well. Afterwards I heard my instructor's voice, over and over again: "With the throttle you should be just as considerate as with your girlfriend. Never treat the throttle roughly." Procedures have to be practiced until one can carry them out when rattled.

Brown pelican (*Pelecanus occidentalis*): W = 30 N, S = 0.5 m^2, b = 2.20 m.

Like good Dutch citizens, my family and I made sure that we had our ice skates with us when we moved to the United States in 1965. That winter, after a couple of nights of hard frost, we drove up into the hills to the small lake behind Whipple Dam in Centre County, Pennsylvania. We heard people whispering to one another as we tied our skates to our shoes. Our wooden slats and leather straps must have seemed like poor substitutes for decent skates. But the whispers faded when people saw that even the strongest teenager on hockey skates could not keep pace with me.

The transmission of your car and that of your bicycle consist of gears that keep the engine's revolutions and your pedaling rate within limits at high speeds. For the engine this is primarily a matter of fuel economy, but for your legs it is mainly a question of endurance. Muscles at work convert glucose into lactic acid; if the acid cannot be eliminated quickly enough, muscle power drops precipitously. Leg muscles are powerful (an athlete's one-hour maximum is about 200 watts), but only if the frequency of the motion remains within limits. A bicycle has a complex set of gears to make this possible, but on speed skates such contraptions are not necessary.

The art of skating produces a fully automatic transmission at absolutely no cost. As you push off, the track of your skate describes a small angle in relation to your direction of travel (figure 11). That angle can be changed. As you increase your forward velocity V, the tracking angle i becomes smaller. You do that automatically. Your legs want to keep the lateral speed w at an acceptable level. This keeps the frequency of your leg movements within limits, thus preventing your muscles from becoming saturated with lactic acid.

The geometry of figure 11 has consequences for the triangle of forces, for the relation between the forward speed V and the lateral speed w, and for your energy budget.

Since ice skates have hardly any friction in the direction of their motion, the force K that you exert with your leg is perpendicular to

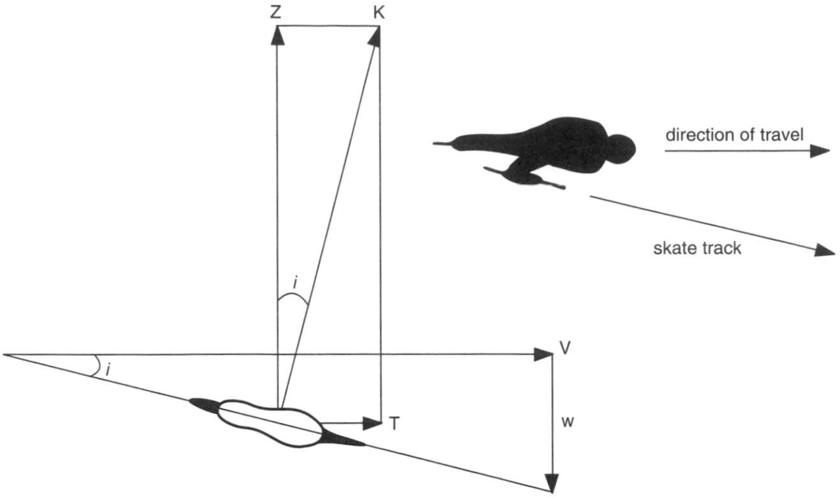

Figure 11 Force and speed for the right-side skating stroke, seen from above. The force K between the ice and the skate is perpendicular to the skate track, becasue the friction of the skate on the ice is extremely small. The motion of the leg generates not only a large sideways force (Z) but also a forward force (the thrust, T). Because the force triangle has the same proportions as the speed triangle, the ratio T/Z is equal to the ratio w/V between the sideways motion of the skate and the forward motion V of the skater.

the skate track. Because the skate track is at an angle i to the direction of travel, the force K has both a lateral component Z and a forward component T (thrust, as before). This is as it should be: without thrust you cannot overcome the aerodynamic resistance.

The force triangle KZT leads to a large lateral force Z and a small thrust T. Because K is perpendicular to the skate track, the angle i between Z and K is identical to the angle i between the direction of travel and the skate track. Hence (this seems trivial, but it is crucial), the proportions within the force triangle are identical to those within the speed triangle. In particular, the ratio between the small thrust T and the large lateral force Z equals the ratio between the low lateral velocity w of the leg and the high forward speed V of the skater:

$$\frac{T}{Z} = \frac{w}{V} . \tag{12}$$

If we write this formula somewhat differently, its elegance is even more striking. Multiplying both sides by V and by Z in order to clear the denominators, we get

$$TV = Zw. \tag{13}$$

The left-hand side of this relation is clearly the power P that is needed for propulsion, just as it is for birds, cars, and airplanes (chapter 2). Power equals force times speed; forward force T times forward speed V equals propulsive power P.

But the right-hand side of equation 13 represents some kind of power, too. The lateral force Z exerted by your legs times the lateral speed of the skate strokes equals the power supplied by your legs. Thus, equation 13 states that the work performed by your muscles during the lateral movement of your legs is converted *without any loss* into the power $P = TV$ needed for propulsion. This is the basis for your "free gearbox": 100 percent of the work done by the large force Z at the small speed w is converted into the work done by the small force T at the high forward speed V. To increase V, you simply push off harder to increase Z without having to increase the frequency of your leg strokes.

The Art of Flapping

A bird in flapping flight faces essentially the same problem as a skater: because flapping its wings too rapidly as its speed increases will cause a buildup of lactic acid in its muscles, the bird must find a way of flapping that limits the frequency of its wing strokes.

The image of flapping that many people have in their heads is one of ducks and geese rowing through the air with their wings. Close observation, however, reveals a very different sort of motion. On the downstroke a bird's wings move slightly *forward*. Rowing with backward strokes would not work. Just imagine moving at 50 miles per hour and having to push your wings back at an even higher rate in order to propel yourself through the air. You would need to maintain a ridiculously high wingbeat rate.

Fortunately, birds are much smarter than that. The flapping of their wings is like a skater's strokes. The only difference is that the plane of action is rotated 90° (figure 12). The downstroke of the wing must generate both lift and thrust. That can be arranged. Because the aerodynamic drag on the wing itself is relatively small, the aerodynamic force K on the wing is almost perpendicular to its direction of motion. As the wing moves down, the force K has not only a vertical component (L), which supplies the lift needed to keep the bird aloft, but also a forward component (T), which provides the required thrust.

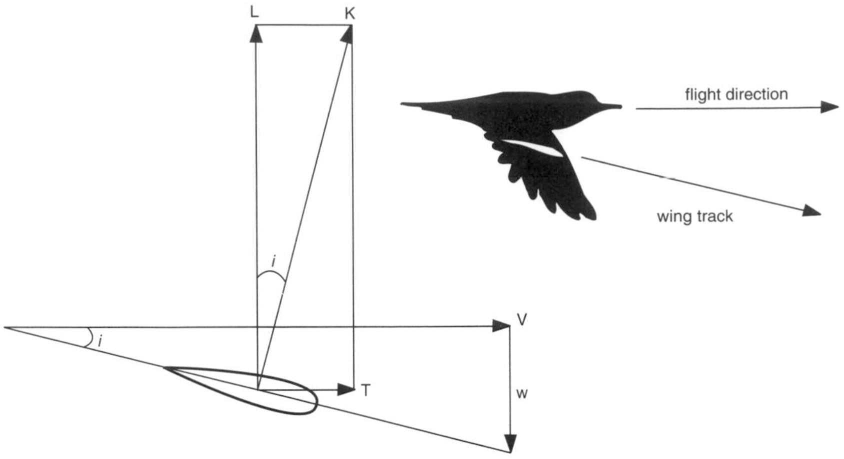

Figure 12 Force and speed for the downstroke of a wing, seen from the side. The aerodynamic force, *K*, is practically perpendicular to the track of the wing, because the air drag on the wing is quite small. The downstroke generates not only lift (*L*) but also thrust (*T*). Because the force and speed triangles have the same proportions, the ratio *T/L* is equal to the ratio *w/V* between the downward speed of the wing and the forward speed of the bird.

The similarity between figures 11 and 12 probably did not escape your attention. The proportions within the force triangle *KTL* are identical to those within the speed triangle, of which *V* and *w* are (respectively) the horizontal and the vertical component. Therefore, the ratio *T/L* between thrust and lift equals the ratio *w/V* between the downward speed of the wing and the forward speed of the bird:

$$\frac{T}{L} = \frac{w}{V}. \tag{14}$$

If we treat this in the same way as equation 12, to ensure that all quantities wind up in the numerator, we obtain

$$TV = Lw. \tag{15}$$

The significance of this is exactly the same as that of equation 13. The product of *T* and *V* is the power *P* required for propulsion. That power is supplied, virtually without loss, by the large force *L* operating at the small downward speed *w*. Lift is necessary to overcome gravity and keep the bird in the air, but as the wing moves down the lift force also generates power. This power, the product of *L* and *w*, is transmitted in its entirety to the propulsive effort. Power, which equals force times speed, can apparently be converted at will from large force times small speed into small force times large speed. As

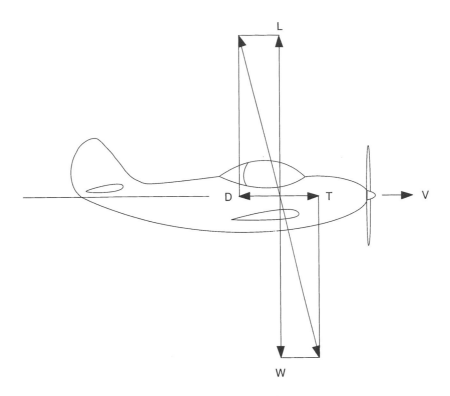

Figure 13 The force balance in horizontal flight. The weight W is balanced by the aerodynamic lift L, and the aerodynamic drag D is overcome by the thrust T. The airspeed is V, and the product of T and V is the work that has to be performed each second (that is, the power required) in order to maintain the force balance. Most of the time, D is much smaller than L. The smaller D/L, the better.

long as a bird can maintain a small angle between the wing stroke and the direction of flight, it can keep the wing-beat frequency down.

Horizontal Flight

The analogy between skating and flapping flight is useful for understanding the way birds and airplanes glide. There, too, the proportions within the force triangle are equal to those within the speed triangle, and there, too, both triangles are slender. But first we must recall the frame of reference developed in chapter 2. The quickest way to do that is to look at the force balance in horizontal powered flight (figure 13). The weight W is kept in balance by the lift L; the aerodynamic drag D is overcome by the thrust T. In horizontal flight at constant speed, therefore, $L = W$ and $T = D$.

Power equals force times speed; thus, the relation between thrust T and power P needed to maintain horizontal flight is given by $P = TV$, just as in skating or in flapping. We can better judge the energy requirements of anything that flies by computing the specific energy consumption $E = P/WV$, which was introduced in chapter 2. The force balance in figure 13 allows us to transform E in various ways. With the aid of $P = TV$ we obtain

$$E = \frac{P}{WV} = \frac{T}{W} = \frac{D}{L}. \tag{16}$$

The ratio D/L between drag and lift determines the specific energy requirements. Tucker's budgy achieved a minimum of 0.22 at a speed of 11 meters per second (25 miles per hour). For parawings and most small birds, the value of D/L is comparable. Seagulls ($D/L = 0.09$), commercial jets ($D/L = 0.07$), and albatrosses ($D/L = 0.05$) are much more economical. Extremely low values of D/L are obtained by state-of-the-art soaring planes, which easily manage $D/L = 0.025$ and in some cases even $D/L < 0.02$. (All these values are minima; at speeds above or below the optimum value the specific energy consumption is higher.)

If you want to save energy in flight, you have to minimize D/L. In flight performance the ratio between drag and lift is a measure of aerodynamic quality. It is inconvenient, however, that this number decreases as the aerodynamic quality is improved. It would be better if we could arrange matters in such a way that the quality number increases as the energy needs come down. And that is quite easy: just turn D/L upside down. The quantity L/D, to which French engineers have given the beautiful and appropriate name *finesse,* has the very properties we desire:

$$\frac{L}{D} = \frac{1}{E} = F. \tag{17}$$

It is a pity that aeronautical engineers in most other countries have given F such unimaginative names. The Dutch and the Germans speak of "glide number" (correct but dull). Anglo-Saxon engineers use "glide ratio" (equally correct and equally dull).

The best finesse a budgerigar achieves is $1/0.22 = 4.5$. The finesse of a wandering albatross is about 20, and that of a Boeing 747 at cruising speed is about 15. The big jet can do somewhat better ($F = 18$) when it flies slower, but its engines are less efficient at lower speeds. Advanced sailplanes achieve $F = 40$ with ease; some reach

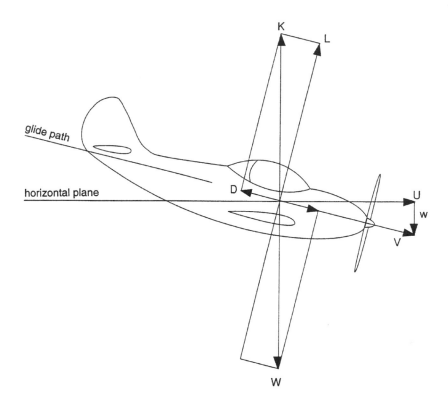

Figure 14 The force balance in gliding flight. With the throttle closed, the thrust equals zero. The drag D now must be overcome by the component of the weight W that is directed along the glide path. Again the force triangle has the same proportions as the speed triangle; hence, D/W equals w/V. (w is the rate of descent and V is the airspeed, now along the glide path). The work performed per second by the weight W is equal to wW, but also to DV.

60. The finesse of a bird or an airplane can be enhanced by slender wings and a smooth, streamlined body.

The Subtleties of Gliding

In the absence of thrust, an airplane cannot maintain the balance of forces required for horizontal flight; it will inevitably lose altitude. When an airplane descends, a new balance is obtained, the component of the weight W directed along the glide path becoming equal to the drag D (figure 14). As with a bicyclist freewheeling downhill, gravity takes over now. The lift L and the drag D constitute the aerodynamic force K that balances the weight W. And, just as in skating and wing flapping, the force triangle is slender: in most cases D is much smaller than L.

The force and speed triangles in figure 14 have the same proportions. Hence, the ratio between the rate of descent w and the airspeed V equals the ratio between the drag D and the weight W:

$$\frac{w}{V} = \frac{D}{W}. \tag{18}$$

As with skate strokes and wingbeats, this can be put in a form that clarifies the energy budget of gliding. Multiplying equation 18 both by W and by V, we obtain

$$Ww = DV. \tag{19}$$

The power $P = DV$ required to overcome the aerodynamic drag is apparently supplied by the large force W acting at the small downward speed w. Power equals force times speed—in this case, the force of gravity times the rate of descent.

The proportionality between the force and speed triangles also provides useful information on the distance that can be covered in a glide. If we call the horizontal component of the airspeed U, the ratio between it and the rate of descent w must be equal to the ratio between the lift L and the drag D. But the ratio L/D is what we have called the finesse, F. What we discover, then, is that the finesse (and only the finesse, since no other quantity is involved) determines how many meters a gliding bird or plane can travel for each meter it descends:

$$F = \frac{L}{D} = \frac{U}{w}. \tag{20}$$

It is for this reason that aeronautical engineers speak of "glide ratio" when they talk about the finesse of an airplane. A jetliner has a glide ratio of about 15, which means that, should all of its engines fail at an altitude of 10 kilometers (33,000 feet), the plane can remain airborne for another 150 kilometers (90 miles). The airliner bringing you to Amsterdam from the United States starts its descent over the British Isles, well before crossing the North Sea. For the same reason, almost half of the flying time on short hops (such as from Chicago to Detroit) consists of descending flight. A jet can easily reach an alternate airport in case of engine failure over Europe or the continental United States; at almost any point there are several airports within 100 miles.

A glider with $F = 40$ can travel 40 feet per foot of altitude lost. At a distance of 4000 feet from its landing spot, the glider needs to be only

Black-browed albatross (*Diomeda melanophris*): W = 38 N, S = 0.32 m^2, b = 2.20 m.

100 feet above the ground. Since that is barely above the treetops, this cannot be regarded as a safe approach procedure. A pilot must be able to see where he or she is going, and the view should not be obstructed by trees or apartment buildings. For this reason all sailplanes are fitted with air brakes (also known as spoilers). With spoilers extended, a sailplane can descend steeply. During the final approach, just before touchdown, F = 10 is more than adequate. (The spoilers on some automobiles are meant not to increase drag but to spoil unwanted aerodynamic lift and keep the wheels in contact with the road at high speeds. But those spoilers are not effective at legal highway speeds; the aerodynamic forces are simply not large enough.)

In a glide, the power $P = DV$ needed to overcome the drag D is supplied by gravity: $P = Ww$ (equation 19). But this means that the rate of descent w can be used as a measure for the engine or muscle power that a bird or a plane must have available to stay aloft:

$$w = \frac{P}{W}. \tag{21}$$

The rate of descent in gliding is equal to the ratio between the power P needed to maintain horizontal flight and the weight W (the "power loading," P/W). The significance of this can be grasped easily by recalling some data from chapter 2. On long flights the pectoral muscles of a bird supply about 100 watts per kilogram of muscle mass. But the flight muscles account for about 20 percent of a bird's mass. The flight muscles therefore supply about 20 watts per kilogram of body mass, which amounts to 2 watts per newton of overall

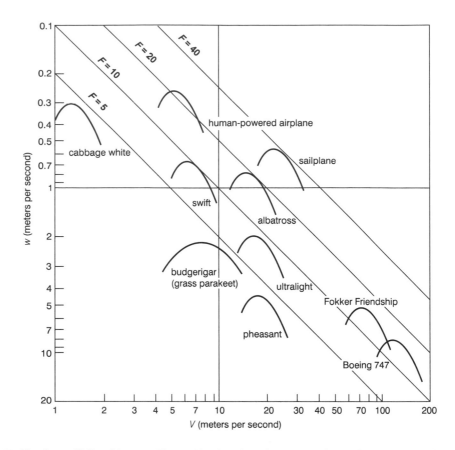

Figure 15 The Great Gliding Diagram. Airspeed is plotted on the horizontal axis. Rate of descent is plotted along the vertical axis, downward. The diagonals running from top left to bottom right are lines of constant finesse. The practical soaring limit, 1 meter per second, is indicated by the horizontal line.

weight. As table 2 shows, watts per newton are meters per second. Equation 21 states that birds with rates of descent greater than 2 meters per second (400 feet per minute) do not possess muscles strong enough to keep them aloft for any length of time.

The Great Gliding Diagram

According to equation 21, we must look not only at the finesse but also at the rate of descent when we want to judge the performance of birds and airplanes. We can do so in an orderly way by plotting airspeed against rate of descent (figure 15). The truism about one picture's being worth a thousand words isn't universally valid, but it certainly holds here. In figure 15—again a proportional diagram,

relative proportions remain the same all over—the per-
aracteristics of insects, birds, and planes can be com-
raightforward way.

great deal to be learned from figure 15. First, let's look at
A pheasant descends at more than 4 meters per sec-
ie maximum we just calculated. A pheasant cannot fly
a minute or so; a quick dash to escape a fox or a
all it can manage. A budgerigar, with a rate of descent
r second, is capable of continuous flight, but without
pare. A larger bird with the same poor value of F (such
would not have lasted long in Tucker's wind tunnel.
fly all over the place. Its continuous power rating is 2
n, but it needs only 0.7 watt per newton to maintain
(figure 15). Hence, it has 1.3 watts per newton to
s that it can climb 1.3 meters per second (roughly 3
without much effort. And if it really wants to exert
conds, its muscles may be able to produce 4 times
or 8 watts per newton. With only 0.7 watt per
maintain altitude, the remaining power can be
3 meters per second (1500 feet per minute). That is
Bonanza. Its speed is equally impressive. With a
rating of 2 watts per newton and a finesse of 10, its
speed is 20 meters per second (45 miles per
ith 4 watts per newton and a finesse of only 7
have to be folded far back), a swift achieves 28
53 miles per hour). The impression of speed is
he human eye judges the speed of a flying object
t its maximum cruising speed, a swift travels 48
l; a cruising 747 manages only 4.

ing flight is
per second
r. For a bird
of the main
100 newtons
apping flight
that do need
But oversize
ncidence that

An aviation
watt of takeoff
ng flight. The
ilogram, or 50
he engine(s) is
10 watts per
descent (equa-
e $F = 12$ as a
sing speed 120
to figure 2, the
r square meter.
about 10^6 new-
with his Spruce
xample, one the
t manufacturers

shows, larger birds must fly faster. With the
ality, which is to say at the same value of
slide along the diagonals toward the lower
gure 15 as they become heavier. Hence, they
more power to remain airborne. Inevitably
when the flight muscles are no longer strong
nit, as we did above, at 2 watts per newton,
of descent of 2 meters per second, and if we
t attainable practical value for finesse, then
cruising speed becomes $2 \times 12 = 12$ meters
hour). Thanks to their extremely slender

Whooping crane (*Grus americana*).

wings, albatrosses achieve $F = 20$, but continuous flapp
beyond their capabilities. A cruising speed of 24 meter
demands a wing loading of 220 newtons per square met
of average proportions (i.e., one that stays in the vicinit
diagonal in figure 2), the corresponding weight is about
(10 kilograms; 22 pounds). Birds that wish to maintain f
for extended periods should not exceed this limit. Those
oversize wings to reduce their power requirements.
wings are ill suited for continuous flapping. It is no coi
most of the very large birds are specialized in soaring,

Piston-engine airplanes have a similar upper limit
gasoline engine weighs roughly 1 kilogram per kilov
power, and about half of that is available in cruisi
specific cruising power, therefore, is 500 watts per k
watts per newton. A practical limit for the weight of
20 percent of the overall weight. This gives P/W =
newton of aircraft weight. The corresponding rate of
tion 21) is 10 meters per second. Again we choos
representative value of the finesse, making the crui
meters per second (270 miles per hour). According
wing loading should then be about 5000 newtons p
An average airplane with that wing loading weighs
tons, or 100 tons. As Howard Hughes discovered
Goose, a piston-engine airplane bigger than that (for
size of a 747) is an untenable proposition. Aircra

converted to jet engines as soon as they could because jet engines have much better power-to-weight ratios than piston engines.

If a bird or a plane wants to stay aloft effortlessly for extended periods, its rate of descent must be less than the rate of ascent in the air along mountain ridges or in convective thermals. A practical limit is 1 meter per second (see figure 15). Budgerigars, chickens, partridges, pheasants, and the like do not need to bother about learning to soar; they descend far too quickly. Modern gliders achieve a rate of descent as small as 0.6 meter per second (120 feet per minute). They can do this because they have extremely slender wings and carefully maintained, smoothly polished skins. Albatrosses are similarly specialized: their minimum rate of descent is 0.8 meter per second (figure 15). The great soaring birds of prey, such as the golden eagle and the California condor, have solved the problem in a different way: they have developed relatively large and broad wings, which give them low cruising speeds. Their rates of descent are within the soaring limit of 1 meter per second, though they do not come anywhere near the albatross in finesse.

Butterflies can soar without worrying much about aerodynamic quality ($F = 4$). Their wing loadings are so low that they can descend at a rate of 0.3 meter per second. But the real maverick in figure 15 is the same one that popped up in figure 2: the human-powered airplane. With the pilot on board, a human-powered aircraft weighs about 100 kilograms, yet its rate of descent is less than that of a 0.15-gram cabbage white.

Induced Drag

It is quite strange that flying should be uneconomical at low speeds. In all other forms of locomotion (e.g., swimming, bicycling, driving) the drag increases as the square of the speed: twice as fast means 4

Emperor moth (*Saturnia pyri*): W = 0.02 N, S = 0.004 m^2, b = 0.1 m.

Brown pelican (*Pelecanus occidentalis*): W = 30 N, S = 0.5 m^2, b = 2.2 m.

times as much drag. But apparently everything that flies can choose an optimum speed, and that optimum need not be unfavorable at high speeds. Though a Boeing 747 flies much faster than a swift, both have about the same finesse. In other words, as a fraction of its weight the drag of a 747 is no larger than that of a swift. Evidently there must be a drag component that *decreases* as the speed goes up.

To stay aloft, a bird must give the air flowing around its wings a downward impulse of sufficient strength to counteract the force of gravity. According to Newton's Second Law of Motion, the force generated by this "momentum transfer" is equal to the product of the downward velocity w imparted to the air and the mass flux q of the air flowing around the wings. The force K thus can be written as

$$K = wq. \tag{22}$$

How much air flows around the wings? If we denote the air density by d, the airspeed by V, and the wingspan by b, we can estimate the mass flux as

$$q = dVb^2. \tag{23}$$

This estimate is not as straightforward as it seems. Although the product of the density d (kilograms per cubic meter) and the volume flux Vb^2 (cubic meters per second) is indeed a mass flux (kilograms of air per second), it is not at all obvious that the square of the wingspan b should appear in the formula. The wing area S would also appear to be a suitable candidate.

Equation 23 states, in effect, that wings have a large radius of influence as far as momentum transfer is concerned. Wings affect the surrounding air not just from wingtip to wingtip, but also above and below them over distances comparable to the wingspan b. Wings have a highly effective way of transferring momentum to the air flowing around them.

Now we can put two and two together. When we combine equations 22 and 23, and take into account that the force K is intended to keep a bird of weight W airborne, we obtain

$$W = dwVb^2. \tag{24}$$

This implies that the downward velocity imparted to the air, w, must be given by

$$w = \frac{W}{dVb^2}. \tag{25}$$

This brings us back to known ground. We have seen time and again that power equals force times speed. That must apply in this case, too. In order to generate an upward force of sufficient strength to keep a bird of weight W aloft, work must be performed at a rate given by

$$Ww = P_i = \frac{W^2}{dVb^2}. \tag{26}$$

This "induced power" must be supplied by the flight muscles. The momentum transfer to the surrounding air is equivalent to an increase in aerodynamic resistance. According to the familiar recipe $P = DV$, the "induced drag" D_i must be equal to the ratio between the induced power P_i and the airspeed V:

$$D_i = \frac{P_i}{V} = \frac{W^2}{dV^2b^2}. \tag{27}$$

This is no minor affair. The square of the airspeed appears in the denominator, not in the numerator. When the airspeed increases, the induced drag decreases in a hurry. When a bird flies twice as fast as its design speed, its induced drag decreases by a factor of 4. But the other side of this coin is devastating: when the speed is halved, the induced drag becomes 4 times as large as is necessary. This is why everything that flies is uneconomical at low speeds, and this is also why the most economical speeds of birds and planes are relatively high.

Equation 27 has another surprise in store. It is not only the square of V that appears in the denominator; the square of the wingspan b is there as well. When you manage to double the wingspan, you are rewarded by a fourfold reduction in the induced drag. To conclude that this is why birds have wings would be only a slight exaggeration.

You could also generate sufficient lift by surfing on your stomach, which is how you fly when you are sky diving. But then you also generate outrageously high induced drag, giving your flight an abominally low finesse. If you want to fly economically, your wings must be slender.

What is a good way to measure the slenderness of wings? Slenderness is the ratio between length and width. But the width of a wing does not remain constant along the span; it usually tapers off toward the tip. For this reason, aeronautical engineers use the ratio b^2/S. The wing area S is equal to the product of the wingspan (from tip to tip) and the average width of the wing. Thus, the ratio b^2/S is equal to the ratio between span and average width. That is exactly what we were looking for. In the English-language technical literature, b^2/S is called the "aspect ratio" and is assigned the symbol A:

$$A = \frac{b^2}{S}.$$
(28)

Dutch engineers use "slenderness" instead, conjuring up shapeliness, grace, elegance, and refinement. This is most appropriate, as we shall see in a moment.

Using equation 28, we can rewrite equation 27 for the induced drag as

$$D_i = \frac{W^2}{dV^2 SA}.$$
(29)

But this contains the familiar combination dV^2S, which we first met in chapter 1. Equation 1 gives a rule of thumb for the relation between the weight W and the cruising speed V:

$$W = 0.3dV^2S.$$

Using this to remove dV^2S from equation 29, and dividing by W once more, we obtain

$$\frac{D_i}{W} = \frac{0.3}{A}.$$
(30)

Like equation 1, this is a rule of thumb, and we cannot expect great accuracy when using it. The coefficient that has been pegged at 0.3 in equation 1 can actually vary between 0.2 and 0.4, the exact choice depending on how an airplane's design is optimized. In the case of jetliners, the coefficient is rather small because jet engines are more economical at higher speeds. But such details don't change the over-

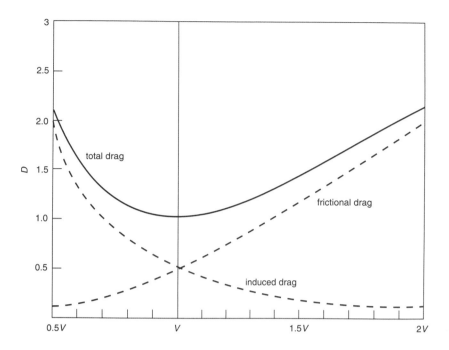

Figure 16 The aerodynamic drag of a bird or an airplane at various speeds. The frictional drag increases as the speed goes up, but the induced drag decreases. The total drag is smallest when the two contributions to the drag are equal; hence, every flying animal or flying machine has an optimum cruising speed.

all conclusion that slender wings have low induced drag. A few examples will make the picture clear. A herring gull has fairly slender wings: $A = 10$. According to equation 30, its value of D_i/W is equal to 0.03. If a herring gull experienced induced drag only, and did not have to worry about the frictional resistance of its body and wings, its finesse would be about 30 ($F = W/D = 1/0.03$). A crow, however, has fairly stubby wings: $A = 5$. It would have a finesse of 15 if it had to cope only with induced drag. In reality both birds have to overcome aerodynamic friction as well. Their true finesses, therefore, are substantially lower: 11 for the gull, 5 for the crow.

Equation 30 states that a large aspect ratio boosts the glide ratio. This connection provides the true justification for the terms "slenderness" and "finesse." Slender wings make for great finesse.

If we were to continue in this way, we might come close to convincing ourselves that wings should always be as slender as possible. That, however, is not true. Aerodynamic friction plays a role that counterbalances the impact of the induced drag. Induced drag decreases as the square of the airspeed, but frictional drag *increases* at the same rate. The total drag D is the sum of the two; it proves to have a theoretical minimum when the frictional drag is equal to the induced drag (figure 16).

Sloppy interpretation of figure 16 can easily lead to faulty conclusions. For example, if we attempt to reduce the induced drag by increasing the aspect ratio while keeping the wing area fixed, we find that the optimum in figure 16 moves to the left, to a lower cruising speed. But when the velocity decreases, the induced drag tends to increase again (equation 29), partly eliminating the advantage of a higher aspect ratio. Equation 30 does not allow for such subtleties. If we want to increase the aspect ratio without compromising cruising speed, we must also improve the frictional drag. The optimum speed in figure 16 then slides back toward the right. The only way to reduce aerodynamic friction is to improve the streamlining of the body and the wings: sleek curves, smooth surfaces, absolutely no loose or poorly fitting feathers, no dangling legs, everything flush and tight.

Conversely, it makes no sense at all to insist on slender wings when it is impractical to maintain a decent streamline shape. House sparrows and hang gliders have relatively high frictional drag. Their wings must be able to withstand frequent folding and considerable abuse, since accidental collisions with obstacles occur all too often. Under those conditions, smooth streamlining does not have high priority. When the frictional drag in figure 16 is increased, the optimum speed moves to the left again. This not only decreases friction but also increases the induced drag, and that is clearly not a smart way to proceed. It makes more sense to maintain the original design speed and accept that wings need not be slender when the frictional resistance is relatively high.

With these deliberations in mind, it is instructive to study the values of F and A for various birds. A number of examples have been collected in table 5. Such a beautiful collection of data begs for comment, but I will restrict myself to sailplanes here.

There is a significant difference between standard-class and open-class gliders. For good reason, the standard class has a prescribed wingspan: 15 meters. If the span were left to the discretion of the designers, everyone capable of building a wing of greater span would be able to achieve a higher finesse (that is, a better glide ratio). A rigid limit on the wingspan amounts to a firm upper limit on A, and thus the standard class needs no handicap rules. Like an albatross, a standard-class glider has an aspect ratio of about 20.

Table 5 Aspect ratio A and finesse F for various birds and airplanes. The values of A have been calculated from $A = b^2/S$; the values of F have been measured or estimated.

	W (N)	S (m^2)	b (m)	A	F
House sparrow	0.28	0.009	0.23	6	4
Swift	0.36	0.016	0.42	11	10
Common tern	1.2	0.056	0.83	12	12
Kestrel (sparrow hawk)	1.8	0.06	0.74	9	9
Carrion crow	5.5	0.12	0.78	5	5
Common buzzard	8.0	0.22	1.25	7	10
Peregrine falcon	8.1	0.13	1.06	9	10
Herring gull	12	0.21	1.43	10	11
Heron	14	0.36	1.73	8	9
White stork	34	0.50	2.00	8	10
Wandering albatross	85	0.62	3.40	19	20
Hang glider	1000	15	10	7	8
Parawing	1000	25	8	2.6	4
Powered parawing	1700	35	10	2.7	4
Ultralight (microlight)	2000	15	10	7	8
Sailplanes					
standard class	3500	10.5	15	21	40
open class	5500	16.3	25	38	60
Fokker F-50	19×10^4	70	29	12	16
Boeing 747	36×10^5	511	60	7	15

In the open class, designers attempt to achieve extremely high values of A. This is no minor undertaking. Aspect ratios as high as 40 make sense only when the skin of the wings, the tail, and the body is extremely smooth, with every seam or crack securely taped. Besides that, the wing spars must withstand enormous bending forces with little flexing, and this makes for a substantial structural weight increase. The empty weight of a standard-class sailplane is about 250 kilograms (550 pounds); that of an open-class sailplane is about 450 kilograms. Nevertheless, it is tempting to increase the aspect ratio of the wings.

The latest version of the Boeing 747 is the 747-400. Its wingspan is 5 meters larger than that of the earlier versions (65 instead of 60 meters), and the wing area is increased from 511 to 530 square meters. The aspect ratio, therefore, is increased from 7 to 8, reducing the induced drag by some 13 percent (equation 29). Since the induced drag of a jetliner is only about one-third of the total drag, a drag reduction of some 4 percent is implied. Boeing's engineers don't exaggerate when they claim a 3 percent reduction in fuel consump-

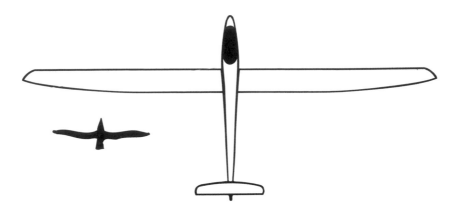

Figure 17 A wandering albatross (*Diomeda exulans*, with b = 3.4 meters) and a standard-class sailplane (Schleicher AK-5, with b = 15 meters) drawn on the same scale. In each case, the wing's aspect ratio is about 20.

tion thanks to the larger wings. The new wingtips also help to increase the effective wingspan somewhat, thus lowering the induced drag even further. They look flashy, too.

Buzzards, eagles, vultures, harriers, and various other birds of prey spend their days soaring around in thermals. A relatively low wing loading keeps a bird's airspeed down, enabling it to achieve a low rate of descent (less than a meter per second) without having to invest in an extremely high finesse. A typical aspect ratio for such a bird is 7. Nevertheless, the finesse of these birds is higher than expected: F = 10. Some investigators have speculated that fully spread primaries (the strong quills on the tip of each wing), with wide gaps between them, produce an effect similar to that of increased wingspan (figure 18). The necessary calculations have been attempted, but a final verdict has not yet been reached.

Hummingbirds and Other Hoverers

Hummingbirds and many insects sip nectar from flowers while hovering in the air. Compared to forward flight, this way of living requires a lot of energy. As might be expected from equation 26, the induced power P_i goes out of control when the forward speed V becomes zero. Some calculations are in order.

For the second time we use the principle that a force equals the product of a velocity imparted by an object and a mass flux. The mass flux generated by the buzzing wings of a hummingbird is about one-fourth of dwb^2, where w is now the downward velocity in the jet of air that keeps the bird aloft. It is useful to compare this with equation 23, where the airspeed V plays the same role as w here. The aerodynamic lift generated by the momentum transfer to the jet is

$$W = 0.25dw^2b^2. \tag{31}$$

Hummingbirds and most insects do not have particularly slender wings: $A = b^2/S$ typically has a value around 6. Substituting this into equation 31, we obtain

$$W = 1.5dw^2S. \tag{32}$$

Again we need equation 1, which reads

$$W = 0.3dV^2S.$$

This allows us to compare the downward velocity w in the jet of air by which the hovering bird or insect keeps itself aloft with the nominal cruising speed V:

$$w = 0.45V. \tag{33}$$

This relation explains why hummingbirds, wasps, bees, and beetles have cruising speeds of roughly 7 meters per second. When we substitute V = 7 meters per second into equation 33, we obtain

$$w = 3 \text{ meters per second.}$$

Marsh harrier (*Circus aeruginosus*): W = 6.8 N, S = 0.22 m^2, b = 1.35 m.

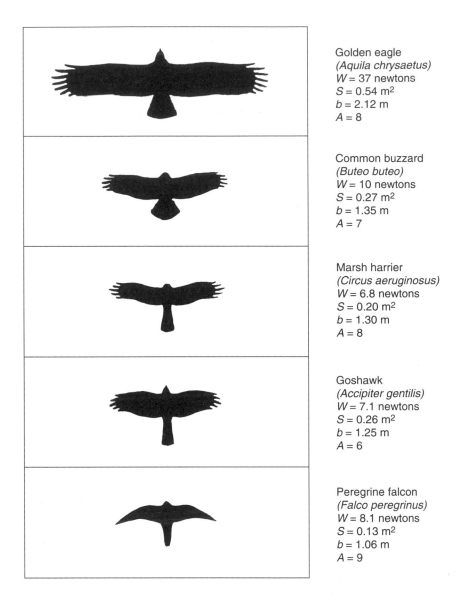

Figure 18 Dimensional data on various birds of prey. All the species in this figure except the peregrine falcon (a high-speed design) have been adapted to slow-speed soaring.

Blue underwing (*Catocala fraxini*): W = 0.012 N, S = 0.0027 m^2, b = 0.08 m.

We have seen this type of expression before. The rate of descent of a gliding bird is a measure of the specific power, P/W, that it needs in horizontal flight (equation 21). For hovering birds and insects, the downward velocity of the airstream generated by the wings plays exactly the same role.

Now it has become easy to draw conclusions. Since the continuous power rating of flight muscles is about 100 watts per kilogram, and since about 30 percent of the overall weight of hovering birds and insects consists of flight muscles, the specific power output is about 30 watts per kilogram of overall weight, or 3 watts per newton. But watts per newton equal meters per second. At full power, therefore, hummingbirds, bees, and the like can generate a jet with a velocity of 3 meters per second to keep them airborne. But that is exactly what we calculated above! In other words, hummingbirds and hovering insects are running at full power continuously. In retrospect, the airspeed of 7 meters per second at which hummingbirds, bees, wasps, and beetles are listed in figure 2 is not primarily a measure of the cruising speed they can maintain. (They have plenty of spare power with which to fly faster if they want to.) Instead, it is a measure of the strength of the jet they can generate beneath their buzzing wings (3 meters per second).

Hummingbirds have evolved to run at full power all the time because the job of transferring momentum to the surrounding air is so strenuous when the forward speed is zero. The same is true for helicopters: they can relax a little only when their forward speed is high enough. In forward flight it is much easier to transfer the required momentum to the air.

How does this work out for larger birds? Isn't it true, for example, that many kinds of ducks are capable of vertical takeoff and landing? The wing loading of a mallard (*Anas platyrhynchos*) is about 120 newtons per square meter. This corresponds to a cruising speed of

Mallard (*Anas platyrhynchos*): W = 11 N, S = 0.093 m², b = 0.9 m.

roughly 18 meters per second. On vertical takeoff a mallard must therefore generate a downward jet with a velocity of 8 meters per second (equation 33: 0.45 × 18 = 8). But that requires 8 watts of takeoff power per newton of weight. That is 4 times the continuous power rating of the flight muscles. A mallard can sustain this much for only a few seconds. After takeoff it shifts into forward flight as soon as it can.

Now we can also understand why the largest of the hummingbirds are much smaller than the largest birds. For hummingbirds the upper limit is about 20 grams, while the largest birds weigh approximately 10 kilograms. Hovering is an uneconomical way of life.

How much fuel does a hummingbird consume? Sugar supplies 14 kilojoules per gram, as does honey. Nectar, half water and half honey, supplies 7 kilojoules of energy per gram. With a metabolic efficiency of 25 percent, the hummingbird's net production of mechanical energy is a little less than 2 kilojoules per gram.

Now we have to calculate the power requirements. The energy transferred to the downward jet of air is Ww joules per second. A 3-gram hummingbird hovering at full power, with w = 3 meters per second, therefore requires a mechanical energy supply of 0.09 joules per second. After all, 0.03 newton times 3 meters per second equals 0.09 watt. Since there are 3600 seconds in an hour, the hummingbird needs a little over 300 joules per hour. Nectar supplies 2000 joules per gram; thus, a gram of nectar suffices for 6 hours of flying.

But this implies that a 3-gram bird consumes its own weight in fuel every 18 hours! One hopes that hummingbirds are permitted some

Puffin (*Fratercula arctica*): W = 2.7 N, S = 0.035 m^2, b = 0.56 m.

rest at night, because a full day's work in the tropical rain forest requires two-thirds of its weight in nectar. This kind of luxury is feasible only in the overwhelming extravagance of a tropical ecosystem, where flowers bloom abundantly throughout the year.

In view of the high fuel consumption of hummingbirds, it is even more amazing that some species migrate over long distances. Every fall the ruby-throated hummingbird travels from the United States to Central America, crossing the Gulf of Mexico on its way, and every spring it retraces its route. I haven't figured out how it manages this, but I am tempted to assume that it can switch from burning sugars to burning fats on long hauls. The journey across the Gulf takes about 30 hours (500 miles at 17 miles per hour). I can't imagine how the hummingbird could perform that feat on sugar water alone, even though it is smart enough to wait for a firm tailwind.

Gossamer Albatross: W = 940 N, S = 70 m^2, b = 29 m.

Among the redeeming qualities of our species is that we play. Indeed, we surround ourselves with toys, and we remain preoccupied with them throughout life.

Flying toys range from paper airplanes, kites (some simple, some exotic), boomerangs, frisbees, aerobees, and spring-powered toy birds to radio-controlled model airplanes to full-scale hot-air balloons, blimps, airships, hang gliders, ultralights and microlights, and sailplanes. There are home-built racers, human-powered airplanes, and even a solar-powered plane.

We display almost inconceivable creativity as we tinker with our playthings. The force of imagination and the passion for experimenting propel us toward outrageous designs and technological advances.

By the middle of the 1970s, sailplanes had achieved a finesse of 40 and a rate of descent of 60 centimeters per second (120 feet per minute). One would think that this would satisfy even the most fanatic of glider pilots. (A lot of work was needed to maintain this kind of performance. Every weekend started with hours of scrubbing and polishing, and after every flight the dead bugs had to be removed from the leading edge of the wings. To keep the finesse from dropping to 20, the wings had to be perfectly smooth.) Designers didn't see much scope for progress at the time: more sophisticated airfoil designs would require expensive wind-tunnel tests and computer simulations, and longer wingspans were not possible with the structural materials then available.

The wind-tunnel problem was not as difficult as it seemed. The smart thing to do when you need sophisticated equipment is to get in touch with professionals. Before you know it, they are as enthusiastic as you are and will use their spare time to do the necessary research. The German Schleicher ASW-22B open-class competition sailplane, for example, achieves a finesse of 60 in part because of a wingtip design that was lovingly perfected in Delft, in the same aerospace

Indoor flying model: W = 0.02 N, S = 70 m^2, b = 29 m.

engineering department where I studied many years ago. It must have cost untold hours to wring that last bit of progress out of a mature technology.

The great breakthrough in sailplane construction came around 1980, when various new materials arrived on the scene: carbon and aramid fibers, expanding foam fillers that made "sandwich" construction possible, and adhesives that could withstand structural stress. Of course, it was some time before appropriate assembly methods were developed, but after that the designers had a field day. In airplane technology, where every ounce of superfluous weight must be avoided, such opportunities are exploited all the way. Boeing's 747-400 saves about 7000 pounds through advanced materials and construction techniques. If a North Atlantic carrier were to use the saved weight to carry an additional pallet of fresh flowers across the ocean, it would earn almost $10,000 extra. But toymakers are just as quick to take advantage of space-age technology. The frame of the Revolution kite is made of "100% aerospace graphite," most kites nowadays have Dacron sails and Kevlar or Dyneema strings, the wing spars of human-powered airplanes are made from aramid fibers, and nearly

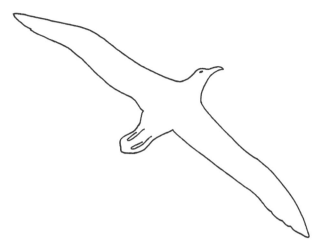

Wandering albatross (*Diomeda exulans*): W = 85 N, S = 0.62 m^2, b = 3.5 m.

all sailplanes are now built with vacuum-formed composite construction techniques, which combine low weight with great rigidity.

Through such advances, today's best competition sailplanes, with a wingspan of 25 meters, achieve a finesse of 60 and a rate of descent of 40 centimeters per second (80 feet per minute). In finesse they beat the albatross, the best nature has to offer, by a factor of 3, and in rate of descent they outperform their nearest avian competitors, swifts and swallows, by a factor of 2, even though these birds have a substantially lower weight, wing loading, and cruising speed.

In fact, open-class competition gliders perform so well that they carry 200 liters of water as ballast. Aeronautical engineers are always keen to save weight, yet here are competition sailplanes taking ballast along! Why? As we saw in chapter 4, the distance covered in gliding is determined by the finesse. In turn, the finesse is determined by streamline shape, surface smoothness, and wing aspect ratio, *not* by wing loading. As the weight of an airplane increases, its speed must increase, but its finesse remains the same. Therefore, if you are in a hurry or if you want to cut your losses in a headwind, you are better off if you are overweight: that increases your cruising speed. The only sacrifice you make is that your rate of descent increases somewhat, but it was extremely low anyway. So, contrary to all aeronautical principles, you make your sailplane heavier than is strictly necessary. And, in order to have the best of both worlds, you arrange it so that you can dump your ballast when the updrafts are weaker than anticipated. Water is useful for this purpose, since at

Questair Venture: W = 8000 N, S = 6.76 m^2, b = 8.40 m, P = 224 kW (300 hp).

worst it might create an unexpected shower for some innocent by-stander.

There is a wide spectrum of extremes in flying playthings. At one end we find indoor flying models, which are designed for extremely low speed. Even with a rather modest finesse (F = 10 at best), such a model achieves a very low rate of descent. Its wing loading is only 0.1 newton per square meter, one-tenth that of a small butterfly. The largest of these models have a 90-centimeter wingspan (3 feet), a wing area of 1000 square centimeters, and a weight of 2 grams (less than a sugar cube). Half of that weight is accounted for by a tightly wound rubber band driving a very large, light, and slow propeller. In covered stadiums these models achieve flight durations up to 45 minutes, traveling only half a meter per second. Once the propeller stops turning, the gossamer contraption loses altitude at a rate of 5 centimeters per second (10 feet per minute). Its rate of descent is one-sixth that of a cabbage white.

At the other end of the scale are sports planes specifically designed for racing. The Questair Venture is fitted with a 224-kilowatt (300-horsepower) Porsche engine. It reaches a top speed of 463 kilometers (290 miles) per hour on wings with a surface area of less than 7 square meters. The engine alone accounts for 30 percent of the takeoff weight. When you hit the throttle for the first time in this machine, you had better take care; it is probably just as temperamental as the late-model Hurricanes and Spitfires of World War II. If you opened up the throttle just a bit too fast in one of those fighters, the propeller's torque would flip you right over.

Whereas the top speed of a 300-horsepower sports car is only 160 miles per hour, a sports plane with the same engine goes nearly twice as fast. If you really get a kick from speed, if it is in your bloodstream, you would do better to take up flying. At least you won't be a menace to others on the freeways.

You don't really need several hundred horsepower to have plenty of weekend flying fun; 30 will suffice. You can see why at any airstrip where ultralights are flown. You will find them there in many shapes and sizes: ragtag contraptions held together by steel cables and covered with spinnaker nylon. They look like hang gliders with tricycle gear and lawnmower engines. If you want to fly a hang glider, you must first find a mountain slope and wait for sufficiently strong winds, but with an ultralight you can come and go as you please, even on the plains and in quiet weather.

Suppose you want to design your own ultralight. Allowing 70 kilograms for your own weight, 40 kilograms for the wings, 30 kilograms for the engine and the propeller, and several tens of kilograms for wires, cables, piping, frame, wheels, and a small fuel tank, you can estimate the total flying weight as 200 kilograms (440 pounds). Let's assume you wish to cruise at 60 kilometers per hour (almost 17 meters per second). The first step in the design process is to consult equation 2, which shows that a wing loading of 106 newtons per square meter is required. From this number and the 2000-newton overall weight, it is a small matter to calculate that the wing area must be 19 square meters (200 square feet). Now move on to figure 15 in chapter 4. A finesse of 8 would seem reasonably conservative; you can't pretend to aim for sophisticated streamlining. With a cruising speed just under 17 meters per second and a finesse of 8, the rate of descent of your ultralight will be a little over 2 meters per second. But the rate of descent tells you how much power you need to keep your weight airborne: $w = P/W$, as equation 21 shows. Since $W = 2000$ newtons and $w = 2$ meters per second, this puts the power required at 4000 watts. But that is not enough. If you want to be able to climb at a rate of 3 meters per second (600 feet per minute), you will need an additional 6000 watts, for a total of 10,000 watts. Ten kilowatts, or 14 horsepower, doesn't seem much. But you haven't yet factored in that you are stuck with a somewhat small and fast

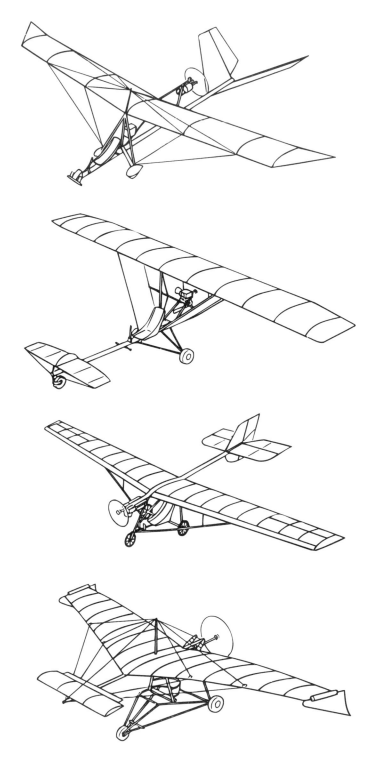

Figure 19   Ultralights (microlights). Source: *Scientific American*, July 1982.

propeller. In this application a fairly large and slow propeller would be best, but you must make do with a small propeller mounted directly to the crankshaft of a fast-running engine. The best you can hope for is a 50 percent propeller efficiency; and this forces you to select a 20-kilowatt (27-horsepower) engine. Now the time has come to sit down at a drawing board and work out the details of your design, making allowances for the unforeseen so that, for example, a wing design that turns out to be slightly heavier than expected will not be disastrous.

For just about anything that flies it is a good idea to maximize the finesse, given all the other design constraints. A kite, however, doesn't really benefit from a finesse higher than 2. An aerodynamically advanced kite with slender wings will float almost straight above your head. But a kite is not stable in that position. It will behave like an errant sailplane, dangling a slack string behind. Since the string has fallen slack, a brisk tug won't help; the kite will float around until it begins to drift sideways. By the time it draws its string taut, the sideways dive of the beautiful design you worked so hard on over the weekend will have become uncontrollable.

A two-string kite can be maneuvered out of danger, and Dacron cloth and carbon-fiber spars can stand a lot more abuse than the Chinese paper and bamboo spars of earlier days. Still, kites perform best when they draw their strings taut. This requires stalled airflow behind the wings—an aerodynamic condition that birds and aeronautical engineers will do anything to avoid.

An aeronautical engineer would also hesitate before selecting a steam engine to power a plane. It is far too heavy, and its thermal efficiency is hopelessly poor. Yet the first powered model airplane in the world was driven by steam. In 1896 the American aviation pioneer Samuel Pierpont Langley flew a powered model airplane across the Potomac River near Washington, a distance of more than two-thirds of a mile. The craft had tandem wings spanning 12 feet and weighed 30 pounds. With its boiler, the steam engine weighed 7 pounds. The power delivered to the propeller was probably only 300 watts, one-tenth the power output of a combustion engine of the same weight, but the event was a significant step forward in the history of aviation. Seven years later, on December 17, 1903, the Wright brothers made their historic flight at Kitty Hawk.

A multiple kite.

Trials with a Paper Airplane

A book on flight would not be complete without a few pages on paper airplanes. Who hasn't played with them at one time or another? Without getting distracted by intricate folding techniques, can you make a paper airplane that concentrates on the art of flying?

All you need is a 4 × 6-inch index card and a large paper clip (figure 20). (A smaller, thinner piece of paper and an ordinary paper clip are suitable, too; there is room enough for experimentation here.)

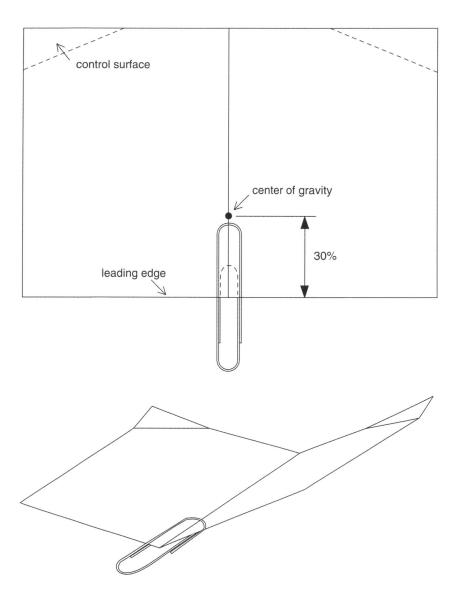

Figure 20 A simple paper airplane. The center of gravity should be located at about 30 percent of the chord (the distance between the leading edge and the trailing edge). The control surfaces in back can be used both as elevators and as ailerons. Once you have adjusted the position of the paper clip for optimum performance, fix it with a piece of tape.

You will see why you need the paper clip as soon as you attempt to launch your model plane. You do that by giving it a push forward, at a speed comparable to walking. But it doesn't want to fly: it starts to tumble backward, like the paddle wheel of a Mississippi steamer in reverse.

The problem is that the center of gravity of the index card lies in the middle of the paper, but the aerodynamic forces are clearly located in front of the center of gravity; otherwise the airplane-to-be wouldn't make those backward somersaults. This problem is solved by shifting the center of gravity of the index card forward—hence the paper clip. The center of gravity of the plane should be located at roughly 30 percent of the distance between the leading and trailing edges of the wing.

Don't make any fold or crease in the index card yet. Position the paper clip carefully, making sure that it sits exactly on the centerline, and resume your flight trials. You will discover soon enough that a paper airplane is extremely sensitive to the exact location of the center of gravity. If you move the paper clip just a little too far forward, the plane dives into the ground; if you slide it back even a fraction of an inch, the plane can't seem to settle down to a smooth and steady glide. As the nose moves up, the plane loses speed until it stalls, causing the nose to drop and the speed to increase. But then the nose moves up again, and the process repeats itself. If the center of gravity is a little too far to the rear, the paper airplane behaves exactly like the immature herring gull mentioned in chapter 3.

Your airplane will be in proper trim when the center of gravity is in the right place. It will then glide smoothly. Knowing that a very high finesse cannot be expected from a simple piece of paper, you may note with satisfaction that your plane glides 4 feet for each foot of height loss. $F = 4$: not bad. Nevertheless, not all is well yet. From time to time your plane starts sliding sideways, slowly at first but gradually faster. Experimenting in a gym or in the stairwell of a large office building, you will find that your plane tends to accelerate into a high-speed "spiral dive." It has a disappointing instability, and it cannot maintain a stable course. But exactly what is going wrong? Since there is still no fold in the index card, your airplane is just a flat piece of paper; it can slide sideways through the air without meeting any resistance. As figure 20 shows, you can correct this by simply making a crease along the centerline. The wingtips are slightly higher now; if the plane tends to slide to the left, the left wing will be pushed up some and the right wing pulled down. The plane rolls to

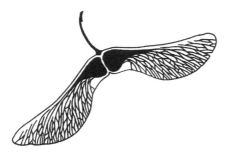

Pair of seeds from Norway Maple (*Acer platanoides*): W = 0.002 N, S = 0.0015 m^2, b = 0.10 m.

the right, starting a right turn to correct for the sideslip to the left, exactly as a bicycle rider counters a gust of crosswind. This is the remedy for the spiral dive. Your plane is now directionally stable.

Once you have caught the spirit of experimentation, you will do well to fix the paper clip with some Scotch tape. It is also a good idea to fold the trailing-edge corners of the wings up a little, so they will serve as control surfaces and make the plane fly slower. Don't overdo it, though; the plane is very sensitive to changes in control-surface angle. With some dexterity and a bit of patience you can make your plane fly near the stalling speed. The control surfaces will also come in handy if the paper clip shifts a little after a crash landing into the furniture. You need not put it back to the original position, since changes in control-surface angle can easily compensate for minor shifts in the center of gravity. An airliner corrects for differences in its load distribution in the same way: if the center of gravity moves forward, the elevators (the movable parts of the tailplane) turn up a few degrees to keep the nose of the airplane up. That is how, notwithstanding variations in loading, an airliner can keep its balance. But there is not much room for error, for the center of gravity must stay within narrow limits.

The control surfaces of your paper airplane can also be used to make it turn. If you want it to turn left, turn the left corner up a little; if you want it to turn right, do the same on the right side. Now you are using the control surfaces as ailerons. Small aileron deflections are needed also to make sure that your plane keeps flying straight if a minor mishap should make it somewhat asymmetrical.

There is one more kind of instability that you may run into: if your paper plane finds it difficult to fly at constant speed, it suffers from the "phugoid." This is best corrected by shifting the center of gravity forward and turning the control surfaces further upward. In any event, you should avoid shifting the center of gravity so far back that the control surfaces must be bent down to maintain trim. That is asking for disaster.

Pedaling Power

People have always dreamed of flying under their own muscle power. In Greek mythology there is the famous story of Daedalus, the inventor in the court of King Minos. Because he had assisted Ariadne in arranging for Theseus to escape, Daedalus was imprisoned in a labyrinth of his own design. But he and his son Icarus managed to flee from Crete by constructing wings made of goose feathers and beeswax. Icarus, the story tells, flew too close to the sun, melting the wax. He crashed and drowned. Daedalus, on the other hand, managed to reach the continent safely. Many centuries later, Leonardo da Vinci toyed with fantasies about helicopters and human-powered airplanes. The flapping wings he tried to design were not a very sound idea: our leg muscles are much more powerful than our arm muscles, not to mention the construction problems encountered when designing oversize flapping wings. But the dreaming continued; a century ago in Germany, Otto Lilienthal experimented with primitive hang gliders until, like Icarus, he crashed, too.

The development of human-powered airplanes began in earnest after World War I. In the years before the war, Professor Ludwig Prandtl of Göttingen, Germany, one of the founders of modern aerodynamics, had systematized the basic principles of flight. The spectacular progress achieved by Prandtl and his colleagues inspired several German universities to include aeronautical engineering in their curriculum, and by 1914 aeronautics had become a popular discipline.

World War I brought the first large-scale use of military airplanes. When the war was over, Germany was prohibited by the Treaty of Versailles from engaging in the design or production of war machinery. As a result, German aeronautical engineers were limited to designing airplanes without engines. But then, as now, designing an actual airplane formed an integral part of the curriculum for senior-year students. Because fighters and bombers were out of the question, German students focused on gliders and human-powered airplanes. To this day, European soaring competitions are dominated by the academic flying teams of German universities: Berlin, Darmstadt, Braunschweig, Karlsruhe, and Stuttgart. Each year, the professors

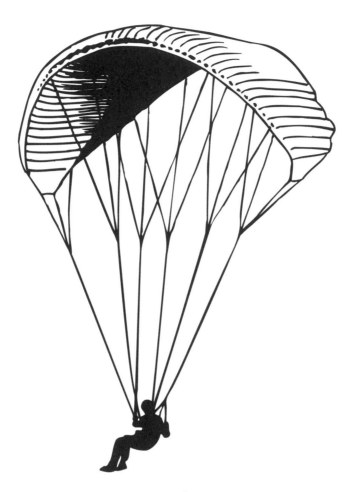

Parawing: W = 1000 N, S = 25 m^2, b = 8 m.

dream up new senior-year design projects in the hope of perfecting their super-sophisticated toys.

In the period between World War I and World War II, the efforts of the German aeronautical engineers resulted in well-engineered gliders with creditable performance for their time; however, apart from a few short hops, their human-powered airplanes were failures. After World War II there was a revival of interest in human-powered airplanes, especially in England, but once again the designers failed to do the simple calculations indicating the extremely low airspeed required for a decent chance of success.

In 1959 the British industrial tycoon Henry Kremer offered a prize of £5000 for the first pilot to complete a figure-eight pattern under his or her own power between two posts half a mile apart. The years passed, but no contender was successful. Kremer gradually increased the stakes, and by 1975 the value of the prize had risen to £50,000.

That was enough to stimulate some serious thinking on the part of Paul McCready, an aeronautical engineer and owner of an environmental consulting firm in California. He started by considering available muscle power. Provided that the effort lasts no longer than several minutes, a well-trained athlete can attain a power output of 250 watts. McCready also realized that the weight of the airplane must be kept as low as possible. A heavy plane must fly fast, thus requiring too much power. McCready decided that the takeoff weight could not exceed 100 kilograms. With a 65-kilogram bicycle racer on the pedals, this left only 35 kilograms for the airplane. And since one can't expect aerodynamic perfection from a lightweight contraption made of corrugated cardboard, piano wire, and Saran Wrap, one can't achieve a very high finesse value. McCready chose $F = 20$, not the $F = 40$ that had become common for sailplanes. Even so, it was going to be an uphill battle.

If the total weight is 100 kilograms and the finesse is 20, then the drag is 5 kilograms, or 50 newtons. It was at this point that McCready made the crucial computation: if you have 250 watts to offer and you have to overcome a resistance of 50 newtons, at what speed can you travel? Since a watt is a newton-meter per second, 250 watts will give you a maximum speed of 5 meters per second against a 50-newton drag (see chapters 2 and 4). Five meters per second, or a little over 10 miles an hour—a speed typical of a teenager on a "granny bike."

To keep 100 kilograms airborne at a speed of 5 meters per second requires extremely large wings. The rule of thumb from chapter 1 is

$$\frac{W}{S} = 0.38 V^2.$$

With $W = 1000$ newtons and $V = 5$ meters per second, the wing area S would have to be more than 100 square meters (1100 square feet)—roughly the floor space of a small two-bedroom apartment. McCready decided that he could achieve the same result with 70 square meters of wing area if the plane were to fly very slowly.

If a plane is to achieve a finesse of 20, its wings must be very slender. According to the data in chapter 4, the aspect ratio must be at least 12 to yield the desired result. On this basis, McCready calculated he would need a wingspan of 30 meters (100 feet).

A hundred feet! That's the height of a ten-story apartment building; that's twice the wingspan of a standard-class glider; that's almost as long as two tractor-trailers; that's the full length of a high school gymnasium! Just think of the design job: 100 feet of wing that must not weigh more than 25 kilograms, because the last 10 kilograms must be reserved for a bicycle frame, pedals, gears, chains, and a propeller. McCready's team succeeded nevertheless, and in the early morning of August 23, 1977, the Kremer Prize was finally won.

With a multitude of problems to overcome, McCready often came close to abandoning the project. Accidental gusts caused several crashes; after a while, all flight trials were conducted before dawn. Fortunately, the primitive construction methods used allowed for quick repairs; McCready says he probably wouldn't have persevered if repairs had taken more time. He was also lucky to have several friends who worked at the Graduate Aeronautical Laboratories of the California Institute of Technology. In the early stages of the project, McCready could not design a suitable propeller. With only one-third of a horsepower available, he could not afford any energy losses. Professor Peter Lissaman helped him out by writing a sophisticated computer program to optimize the propeller design. A few unauthorized weekend runs on Caltech's supercomputer quickly solved the problem.

The board of directors of the giant chemical company DuPont, prepared as ever to support intelligent dreamers, offered to sponsor McCready's project and to supply advanced materials, including Mylar, Kevlar, and carbon fibers. Meanwhile, Peter Lissaman made additional computer calculations, and other associates concocted construction methods that reduced the amount of piano wire and thus the aerodynamic drag. The finesse inched its way upward to 25, while the empty weight of the plane came down at least 6 kilograms. With $W = 940$ newtons and $F = 25$, the drag is not 50 but 38 newtons; at a speed of 5 meters per second the power required is only 190 watts. A professional bicycle racer in good condition can keep that up for several hours. Long-distance trips were coming within reach.

The McCready team was responding to Henry Kremer's latest challenge. Kremer, who evidently loves to kindle fires, had announced a prize of £100,000 for the first human-powered flight across the English Channel from Dover to Calais. That prize was won by Bryan Allen, the same cyclist who had captured the first one. On June 12, 1979, he pedaled the Gossamer Albatross from England to France. Allen had counted on having to pedal for just under 2 hours, but near Cap Griz-Nez he encountered unexpected headwinds. Because of the wind, it took him 2 hours and 45 minutes to reach the other side.

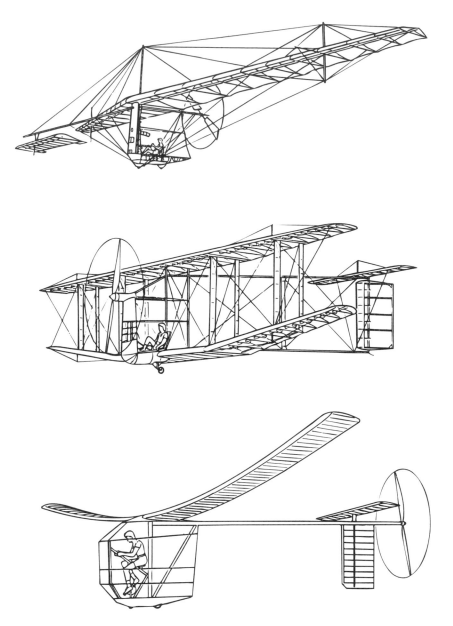

Figure 21 Human-powered airplanes. From top: Gossamer Condor, Chrysalis, and Musculair. The Monarch, built at the Massachusetts Institute of Technology, looks similar to the Muscu-lair. Source: *Scientific American,* November 1985.

Kremer just couldn't stop teasing the human-powered-airplane crowd. McCready's planes flew far too slowly for his taste, making them sensitive to turbulence, gusts, and headwinds. What's the use of a plane that can be flown only at dawn? So he offered yet another prize: £20,000 for the first human-powered airplane to complete a one-mile triangular course within 3 minutes. This would require more than 20 miles per hour. Again engineers and scientists came into action, this time at the Massachusetts Institute of Technology. They used their ingenuity to achieve still greater finesse. Twenty miles per hour is about 10 meters per second; at that speed, a finesse of 33 is needed to bring the drag of a 100-kilogram plane down to 30 newtons. The power required then is 300 watts, a rate a professional bicyclist can maintain for 10 minutes at best. But the tinkering by the MIT crew paid off: in May 1984 their Monarch won the prize.

The dream of Daedalus finally came true in 1988. On April 23 the Greek cyclist Kannellos Kannellopoulos managed to fly due north from Crete to the tiny Cycladic island of Santorini, 120 kilometers away. With the wind at his back, the trip took 4 hours. The finesse had been improved to 40 by this time, and with a cruising speed of 7 meters per second 200 watts was sufficient—too much to demand of an amateur, but within reach of the professional Kannellopoulos. (It took a team of physiologists, ergonomists, and other experts 4 years to select and train the winning contender. Several others were forced to drop out of the competition because of inefficient metabolism or poor muscle discipline. Great strength in itself is not necessarily an asset when it comes to flying.)

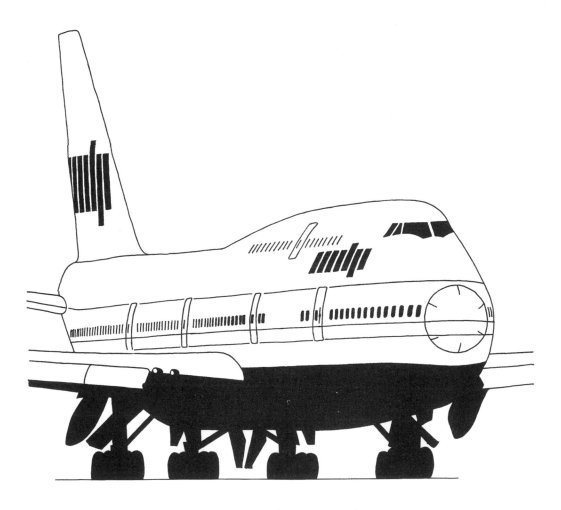

Boeing 747-400: $W = 3.95 \times 10^6$ N, $S = 530$ m^2, $b = 65$ m.

On November 23, 1991, en route to Washington, I was standing in the cockpit of a British Airways Concorde, chatting with the flight engineer. We were flying at Mach 2, an amazing 22 miles a minute. The sun had just risen above the *western* horizon. I scanned the fuel gauges but couldn't find what I was looking for.

"What is your fuel flow?" I asked.

"Twenty tons per hour," the engineer replied.

"That's twice the fuel flow of a 747," I said.

"Yes, but we're going twice as fast."

Twenty tons per hour and $3\frac{1}{2}$ hours in the air amounts to 70 tons of kerosene needed to carry 100 passengers across the ocean! Allowing for some empty seats, this comes to 1000 liters per person. A 747 also consumes 70 tons of fuel between London and Washington, but it carries 350 people and 30 tons of freight along on the trip. The Concorde cannot afford to carry any freight at all. Its belly is full of fuel, yet its range is only half that of a 747. It has to stop in Lisbon to top off its tanks before it can take British jet-setters to Barbados.

I have a love-hate relationship with the Concorde. At my doctoral thesis defense, in 1964, I expounded on the proposition that supersonic airliners would be a step back in the history of aviation. Preliminary designs for the Concorde were on the drawing boards at the time, and the air was thick with feverish speculations on the next great step forward in the triumphal progress of aerospace engineering. One thing was certain: humans needed to fly faster than the speed of sound.

But the finesse of aircraft designed to penetrate the sound barrier is hopelessly low. A Boeing 747 has a finesse of about 15, but the Concorde barely reaches 5. In 1964 there were hopes that this disadvantage could be compensated by the high thermal efficiency of supersonic jet engines, but the advanced by-pass engines of present-day subsonic jetliners, twice as efficient as their early counterparts, easily beat the supersonic competition.

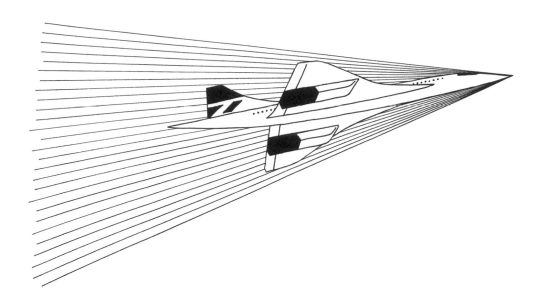

Concorde ($W = 1.8 \times 10^6$ N, $S = 358$ m^2, $b = 25.6$ m), showing shock waves generated.

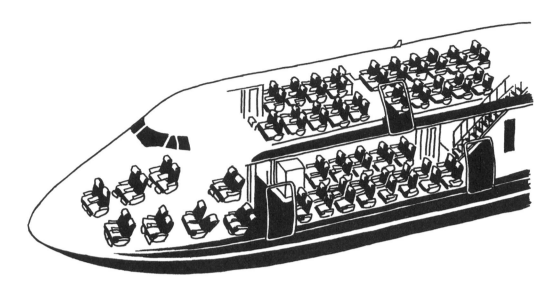

Interior of Boeing 747-400.

An aircraft flying faster than the speed of sound creates shock waves in the air, much like the bow and stern waves of a tugboat crossing a harbor at speed. This is what causes "sonic booms." Creating these waves takes a lot of energy. Because a plane in supersonic flight can't avoid making shock waves, the problem of declining finesse is insoluble. Although Concorde passengers don't notice anything as their plane penetrates the sound barrier, the economic barrier is real enough: if you want to exceed Mach 1, it will cost you 3 times as much. For the aircraft industry this is indeed a step in the wrong direction. Time and again, before aeronautical engineers started dabbling with supersonic flight, they had managed to reach higher speeds at lower costs. The Concorde broke that trend.

The Concorde may be the showpiece of the aviation industry, but the Boeing 747 is one of the great engineering wonders of the world, like the pyramids of Egypt, the Eiffel Tower, or the Panama Canal. The largest airliner in the world, the 747 has been so phenomenally successful that more than 1000 have been built so far. The plane incorporates everything one can reasonably demand of a mode of transportation: reliability, ease of maintenance, productivity, fuel economy, speed, and ample cargo space. For its weight, the 747's wing loading is quite ordinary. However, that weight puts the plane in a class by itself.

A single 747 flying back and forth between Amsterdam and New York produces at least 2 million seat-miles a day, based on 300-plus seats and a one-way distance of 3600 statute miles. Even with maintenance and periodic inspections, a 747 makes more than 300 round trips a year, so its annual production, not including freight, is 600 million seat-miles. Ten 747s match the entire traffic volume of the Netherlands State Railways. Admittedly, Holland is a small country, but nevertheless several hundred coaches and commuter trains are required to achieve that performance. In the United States, an Amtrak train making the two-day run between Chicago and San Francisco with 400 passengers aboard accounts, in theory, for less than half a million seat-miles a day, and in practice delays, maintenance, and time-consuming turnaround procedures cut that in half: the net rate is only 200,000 seat-miles a day. The French high-speed train, the TGV, fares somewhat better. The 250-mile run between Paris and Lyon takes 2 hours and carries 500 people. Making three round trips a day, a TGV, with almost twice as many seats as a 747, produces more than 700,000 seat-miles a day—but that is only one-third of a 747's productivity.

A new 747 costs roughly $150 million. The first owner writes this cost off over 10 years. After 10 years, though, a 747 still has plenty of life in it. Average depreciation and interest over the first 10 years are estimated at $15 million and $7 million per year, respectively, for a total of $22 million per year. One-third of that amount must be earned by carrying freight, leaving roughly $15 million to be recovered by selling 600 million seat-miles. This works out at 2.5 cents per seat-mile. Compare that with your car. The Consumers Union estimates automobile interest and depreciation at about 20 cents per mile. With two people in a car on average, the cost is thus 10 cents a passenger-mile—considerably more expensive than an airplane.

Trains are not very economical on this score, either. Per seat-mile, the direct energy consumption of a train is half that of a car (see chapter 2), but the other costs are disappointing. Railroads must maintain an extensive infrastructure and must invest heavily in their inefficiently utilized rolling stock. In order to break even in their passenger operations, railroads the world over depend on massive state subsidies. As a rule of thumb, the price of a railroad ticket covers only half of the real cost.

It is easy to see why the great ocean liners were doomed as soon as jetliners appeared on the transatlantic market. A ship of some stature

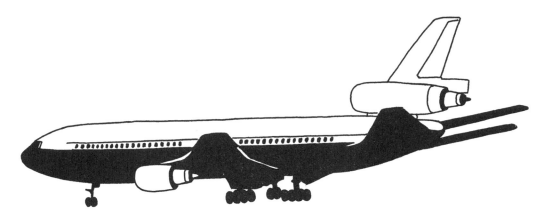

Douglas DC-10: $W = 2.56 \times 10^6$ N, $S = 368$ m^2, $b = 50$ m.

easily costs $500 million. Let's assume that a round trip between Southampton and New York takes 2 weeks; that allows 20 round trips a year if we exclude the winter season. With 1500 passengers on board, an ocean liner produces 200 million passenger-miles a year. If we estimate interest and depreciation optimistically at $50 million a year, the cost comes to 25 cents a passenger-mile—10 times the capital expenses of a 747. Low-speed travel has its advantages, primarily as far as direct propulsion costs are concerned, but it remains unprofitable when you look at total productivity. A slow mode of transportation makes it difficult to recoup one's investment.

The Boeing 747 is also unequalled as a freight carrier. Let's compare freight against passengers. With luggage and meals included, passengers account for an average of 100 kilograms each. On intercontinental flights economy-class passengers pay approximately 10 cents a mile (much less during one of the frequent "fare wars"). Of this, 2 cents is for fuel and 2.5 cents for capital expenses. Converted to weight, the passenger tariff becomes roughly a dollar per ton-mile. But freight doesn't require flight attendants, chairs, pillows, blankets, toilets, kitchens, and meals. It should therefore be possible to offer freight service at approximately half the price per ton-mile. Indeed, the going rate for intercontinental airfreight is about 60 cents per ton-mile, increasing to a dollar per ton-mile for fresh-cut flowers, vegetables, and other perishables, and with additional surcharges for horses, elephants, and day-old chicks.

Imagine you are a fashion buyer at Macy's in New York or Harrods in London, and you need an extra supply of some suddenly hot-sell-

Lockheed Tristar: $W = 2.2 \times 10^6$ N, $S = 320$ m^2, $b = 50$ m.

ing jeans from a supplier in Asia. Now, 2000 pairs of denim trousers, at roughly a pound each, weigh about a ton. A Boeing 747 can carry that order 10,000 miles at a cost of $6000, or $3 per pair. Your department store can recover this expense with ease by increasing the price of the jeans from $49 to $59.

Gladioli from South Africa can't lie flat; they have to be transported upside down to prevent their tips from becoming crooked. Roses from Argentina find their way to American flower shops on the same day. Freesias from Israel are auctioned off at dawn near Amsterdam and are on their way to Chicago before noon. The sky is literally the limit in the variety of products that are flown around the world. Take off-season string beans: 1200 miles between Tunisia and Holland, at a rate of 60 cents a ton-mile, make the transportation cost only 35 cents a pound. In the middle of winter I certainly would not mind paying that extra charge for a special treat.

Though airfreight has a reputation of being expensive, the differences are not as great as one might think. Federal Express and UPS offer surface rates not much below regular airfreight. Indeed, Federal Express owns a fleet of cargo planes, and they wouldn't do that if it weren't profitable. Substantially lower rates are offered by long-distance trucking companies. Bulk shipments of ore and grain by rail are cheaper still, and barges on the Great Lakes charge only 2 cents a ton-mile or even less. For the time being, therefore, one should not expect to find planes being loaded with ore or gravel. Nevertheless, the potential of airfreight should not be underestimated. Once fully depreciated 747s are bought up by the Flying Tigers and other tramp operators, the price may drop to 30 cents a ton-mile. The first 747 entered service with Pan American in 1969; not all that many superannuated 747s are around as yet. But the earlier long-distance airliners, the Douglas DC-8s and Boeing 707s, have long since been bought up by freight operators. Just look around the tarmac the next time you are waiting for a connecting flight at Atlanta or Pittsburgh: the far ends of the apron are cluttered with unmarked airliners painted an unattractive gray, with their windows riveted shut. Before long, the early 747s will start showing up in such fleets. This is just the fate that was foreseen by Juan Trippe, the strong-willed president of Pan Am. In the mid 1960s he negotiated with Boeing for a wide-bodied jet that could easily be converted to a cargo plane after the passenger market was captured by the supersonic jetliners he expected to arrive shortly. Trippe was badly mistaken about the prospects of supersonic flight, but his insistence forced Boeing to design the best airliner ever built.

Like all intercontinental airliners, the Boeing 747 is designed with the North Atlantic in mind. This is the corridor with the most business, the fiercest competition, and the most potential income. By a stroke of luck, it also happens to be the run that provides the most efficient flight schedules. In the 1950s, a propeller plane took an average of 14 hours to make the crossing; a day later it began its return trip. Nowadays a westbound flight takes about 8 hours, and 3 hours later the plane is on its way back, to land in London, Amsterdam, or Frankfurt 7 hours later. Just 18 hours after its departure a 747 is back home, in time for the early cleaning and maintenance shifts. An excellent routine: no clumsy personnel from outside contractors, no overtime payments, and a fixed pattern of home port maintenance running like clockwork. Such a routine helps to keep costs down.

On the longer runs between Europe and the United States there is little time to spare. Rome–New York or Amsterdam–Chicago must be completed within 9 hours, or a fixed schedule with the plane back home every morning becomes impossible. The journey from Rome to New York is 4300 miles—700 miles longer than the Amsterdam–New York run. A travel time of 9 hours demands a cruising speed of at least 500 miles per hour to allow time for taxiing, takeoff, waiting in the holding pattern, and landing.

To keep up the same kind of routine on longer trips, the cruising speed has to become even higher, but then you run up against the speed of sound. Economical flight requires that the cruising speed remain below the speed of sound. Shock waves cost too much fuel. In practice, Mach 0.9 is the limit. If the speed of sound were 500 miles per hour, for example, the maximum cruising speed would be 450 miles per hour. But at that speed you couldn't possibly design a decent transatlantic timetable for long-distance jets, and this would be a major threat to productivity. An airplane must fly in order to earn its keep; every hour on the apron increases the seat-mile costs.

What is the speed of sound, incidentally? You may recall a rule of thumb from childhood: each 5 seconds between a flash of lightning and its accompanying peal of thunder puts a mile between you and the thunderstorm (3 seconds for each kilometer). The speed of sound at sea level is about 760 miles per hour. At 40,000 feet, where the temperature is more than 50°C below zero, it is about 670 miles per hour. To keep below Mach 0.9, a Boeing 747 cannot exceed 600 miles per hour. Its most economical cruising speed is lower yet: 560 miles per hour, or Mach 0.83. This doesn't mean, however, that you will read 560 mph on the airspeed indicator if you glance into the cockpit. The number you will see is 245 knots, which amounts to 283 mph. Why such a low figure? In fact, the airspeed indicator measures not the velocity V but the product dV^2 that we encountered in chapters 1 and 4. It is not possible, however, to measure the air density separately, and the mechanism inside the indicator is calibrated for sea-level density, which is 4 times the density at 40,000 feet. The airspeed indicator is therefore off by half, and the speed really is 565 miles per hour.

Allowing for headwinds and various delays, the 4000-mile trip between Amsterdam and Chicago takes 9 hours. If the trip took any longer, the 747 would have to stay in Chicago overnight. An idle airplane is expensive, especially when it costs more than $20 million a year in interest and depreciation alone. (That's $2500 an hour, or 70 cents a second.)

The speed of sound is high enough to contribute to the profitability of the busiest air-traffic corridor in the world. But all long-distance jetliners profit from this coincidence, not just the 747. Nor need the size of the 747 impress us. No, the 747 is special only because it is a wonderfully logical solution to an uncompromising engineering problem. From the viewpoint of economy of scale, it is a happy coincidence that the optimal solution necessitates a large and heavy airplane; however, as we shall see, size in itself is not part of the argumentation. The size of the 747 is not a matter of choice, even though it may have seemed that way to Mr. Trippe and the Boeing design team.

Requirement 1 An airliner must travel as fast as possible without a major sacrifice in finesse. The higher its speed, the smaller the impact of the capital expenses. If depreciation alone is going to cost you several million dollars a year, you can't afford to drag your feet.

Requirement 2 An airliner must fly below the speed of sound, or it will suffer from the drastic drop in finesse that occurs at supersonic speed. If supersonic flight didn't require so much fuel, it would be great; but it just doesn't work out that way. Mach 0.9 is an absolute maximum.

Comparing these two requirements, we deduce that a cruising speed just below Mach 0.9 is both a minimum and a maximum. We have painted ourselves into a corner, but there is a bonus thrown in: jet engines are more efficient when they fly faster. Speed acts like a turbocompressor, so at Mach 0.9 jet engines enjoy an appreciable amount of turbo boost.

Requirement 3 The colder the air, the better. The efficiency of jet engines improves as the difference between the intake temperature and the combustion temperature increases. Let me remind you of some thermodynamics: the efficiency of converting heat into useful work depends on temperature differences, as was discovered by the French engineer Sadi Carnot (1796–1832).

This fact has far-reaching consequences for jetliners. The coldest air is found in the lower stratosphere, above 10 kilometers (33,000

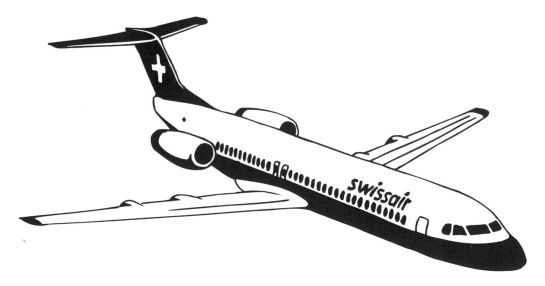

Fokker F-100: $W = 4.3 \times 10^5$ N, $S = 94$ m^2, $b = 28$ m.

feet). The temperature there is about 55°C below zero. Long-distance planes must fly high in order to travel far, but again a bonus is thrown in: clouds and thunderstorms are extremely rare in the stratosphere, so flight schedules can disregard meteorological conditions. In the stratosphere, airplanes fly "above the weather." This is advantageous for passengers too: above the weather there is very little turbulence.

Flying high has yet another advantage. At high altitudes, the air is much less dense than at sea level. In order to remain airborne at 33,000 feet, an airplane needs relatively large wings. Near the ground, these wings permit much lower speeds: the sea-level cruising speed is about half the cruising speed at altitude. If the wings are fitted with extensive flaps and slats, the takeoff and landing speeds are lower yet. This helps to limit the required runway length.

Requirement 4 An airplane shouldn't fly one foot higher than is necessary. Rarefied air requires oversize wings, like those of the American U-2 spy plane. Moreover, jet engines operating in rarefied air suffer respiration problems. If you insist on flying too high, your plane needs not only oversize wings but also oversize engines.

Other design factors may interfere with the fourth requirement. The Concorde, for example, has oversize wings because it has to take off from and land on the same runways as subsonic jetliners. In order to minimize the disadvantages of those wings, the Concorde must fly high in the stratosphere, at 58,000 feet.

Together, the third and fourth requirements suggest 10 kilometers (33,000 feet) as the correct cruising altitude. It does not pay to go higher, because it is just as cold higher up. Conditions are optimal at the tropopause (the boundary between the troposphere and the stratosphere), where the highest density consistent with the quest for low outside temperatures is found. Again we have boxed ourselves in: the cruising altitude is not left to the designer's discretion, but is determined by straightforward engineering logic.

A well-designed jetliner must fly a little slower than Mach 0.9 at a cruising height of 10 kilometers. We choose a speed of 250 meters per second (900 kilometers per hour, 560 mph, Mach 0.83). That is on the safe side of the absolute maximum. Now we can start drawing conclusions. First of all, we need equation 1, which shows how wing loading depends on density and airspeed:

$$\frac{W}{S} = 0.3dV^2.$$

Table 6 Atmospheric data: altitude h in meters, temperature T in degrees Celsius, and air density d in kilograms per cubic meter. Also shown are the ratio V/V_0 between cruising speed at altitude and sea-level cruising speed and the relation (expressed as a percentage) between weight loss and altitude gain for a jetliner starting an intercontinental journey at h = 9000 meters.

h	T	d	d/d_0	V/V_0	% W_0
0	15.0	1.225	1.000	1.00	
1000	8.5	1.112	0.908	1.05	
2000	2.0	1.007	0.822	1.10	
3000	−4.5	0.909	0.742	1.16	
4000	−11.0	0.819	0.669	1.22	
5000	−17.5	0.736	0.601	1.29	
6000	−24.0	0.660	0.539	1.36	
7000	−30.5	0.590	0.481	1.44	
8000	−37.0	0.525	0.429	1.53	
9000	−43.5	0.466	0.381	1.62	100
10000	−50.0	0.413	0.337	1.72	89
11000	−56.5	0.364	0.297	1.83	78
12000	−56.5	0.311	0.254	1.98	67
13000	−56.5	0.266	0.217	2.15	57
14000	−56.5	0.227	0.185	2.32	
15000	−56.5	0.194	0.158	2.52	
16000	−56.5	0.165	0.135	2.72	
17000	−56.5	0.141	0.115	2.95	
18000	−56.5	0.121	0.099	3.18	

At an altitude of 10 kilometers, the air density d is 0.413 kilograms per cubic meter (see table 6). If we substitute d = 0.413 and V = 250 into equation 1, we find that the wing loading of our airplane, W/S, should be 7740 newtons per square meter.

What is the gross weight of a mundane, economical airplane with that wing loading? Both oversize and undersize wings have their drawbacks, so let's stay in line with the main diagonal in figure 2:

$$\frac{W}{S} = 47 \sqrt[3]{W}.$$

The result is $W = 4.47 \times 10^6$ newtons, or 447 tons. The wing area then becomes 578 square meters (about 6200 square feet).

It would have been very easy to manipulate these numbers in a way that yields the precise figures for the 747. The 747-400 has a maximum takeoff weight of 394 tons and a wing area of 530 square meters (5700 square feet). Our calculations would have come close if we had taken into account that the numerical coefficient in equation 1 should be somewhat smaller than 0.3 when computing the cruising speed of a jet, because of the improved engine efficiency at higher speeds. But that would have been nitpicking. Equations 1 and 3 are merely rules of thumb; great precision cannot be expected. Nor is great precision required to realize that the weight of an airplane designed to fly 900 kilometers an hour at an altitude of 10 kilometers should be roughly 400 tons.

Despite my chain of reasoning, you may think that these arguments are simply meant to promote the 747. What about other types of airplane that sell well? Most jetliners are compromises between conflicting design criteria; the Boeing 747 is the only one that obeys ruthless engineering logic. The principles of flight take the decision out of our hands. Yet you shouldn't believe these conclusions unless I support them with further argumentation. The best way to argue the case is to see what happens if we deviate from the rules. After all, the main diagonal in the Great Flight Diagram allows plenty of scope for the ingenuity of aircraft designers. A designer can choose to move to the left or to the right of the main diagonal, giving a plane a lower or a higher wing loading than is typical for its weight.

Need a jetliner weigh 400 tons? Of course not; it is easy enough to design a smaller airplane. But a smaller airplane has a smaller wing loading if it has wings to suit its size, and a lower cruising speed to match. If you nonetheless insist on cruising at 560 miles per hour, you need to fly abnormally high. The Boeing 737 weighs 57 tons; if it had ordinary wings for its size, its wing loading would be 3900 newtons per square meter and its wing area 146 square meters (equation 3). At a design speed of 900 kilometers per hour, or 250 meters per second, the air density would have to be 0.2 kilograms per cubic meter (equation 1). According to table 6, the corresponding cruise altitude is 15 kilometers (49,000 feet). But a short-distance airliner, a flying commuter bus, cannot afford to climb to such an altitude. For this reason the designers of the 737 had to compromise both on cruising speed and on altitude. There was an advantage to be gained with a higher-than-average wing loading (that is, with under-size wings). With a wing size of 105 square meters (see table 7), the

Table 7 Data on popular airliners. Sources: *Jane's All the World's Aircraft, KLM Holland Herald, Martinair Magazine, Transavia Inflight Magazine.*

	Takeoff weight W (tons)	S (m^2)	b (m)	Sea-level thrust T (tons)	Fuel consumption (liters/hour)	Cruising speed V (km/hour)	Range (km)	Seats
Boeing 747-400	395	530	65	4 × 25.7	12300	900	12200	421
Boeing 747-300	378	511	60	4 × 23.8	13600	900	10500	400
Boeing 747-200	352	511	60	4 × 21.3	13900	900	9500	387
Douglas DC-10-30	256	368	50	3 × 23.1	10400	900	9900	248
Airbus A310	139	219	44	2 × 22.7	5500	860	6400	200
Boeing 737-300	57	105	29	2 × 9.1	2700	800	4200	124
Fokker F-100	43	94	28	2 × 6.7	2400	720	1800	101
Fokker F-28	33	79	25	2 × 4.5	2500	680	1700	80

wing loading of the 737 is in fact a little over 5400 newtons per square meter—40 percent higher than average for its size.

Why not produce a 737 with the same wing loading as a 747? The wings would be too small. The wings of a 57-ton plane with a wing loading equal to that of a 747 would be no larger than 77 square meters (830 square feet), which is only half the median value for the plane's weight. A bird with undersize wings has a comparatively fat body, which creates additional air resistance and renders the design less economical than it should be. The 737 would look even more like a puffin than it does already. Moreover, a 737 with a wing loading equal to that of a 747 would need 2-mile runways. It would not be able to land at smaller airports, where the standard runway is only a mile long. The 737 is intended for short distances, and a commuter plane unable to use regional airports makes no economic sense. Although the wing loading of a 737 is above average for its weight class, it is not excessive.

The Boeing 737 is a reasonable compromise between the desire to fly faster than is appropriate for its weight and the price to be paid for

Boeing 737: $W = 5.7 \times 10^5$ N, $S = 105$ m^2, $b = 29$ m.

insisting on doing this. It is a sensible solution to conflicting design specifications, but it remains a compromise.

What about a much larger airplane—a 1000-ton giant with a wing loading of 10,000 newtons per square meter? To fly at a speed of 900 kilometers per hour, it would need to travel at an altitude where the air density is 0.53 kilograms per cubic meter (equation 1). But that corresponds to 8 kilometers (26,000 feet; see table 6), which is not high enough to be above the weather. Also, it is not cold enough at this height: only 37°C below zero, not 56°C below as in the stratosphere. The wing loading of the 1000-ton airplane would have to be reduced in order to reach a cruising height of 10 kilometers. This means that the plane would need oversize wings, making them heavier than necessary and reducing the payload capacity.

To design a 1000-ton plane, therefore, would entail a number of compromises. For one thing, the wing loading could not be higher than that of the 747, because available runways would be too short. The designers would probably also run into trouble with the structural weight. It was possible to design the 747 only after titanium alloys, which are much stronger than the best steel and aluminum alloys, appeared on the market. A yet heavier plane would require even more advanced materials, or would see its payload reduced. For the time being, I don't see any aircraft company daring to build a 1000-ton, 1000-passenger airliner. For as long as I can remember, designers have floated speculations on super-jumbos, but I have never seen more than fancy sketches in trade magazines. This is one field where there are real limits to growth. The Boeing 747 is so sensationally successful that it discourages competition. Its assembly line has been running for 25 years; I am sure it will run another 25.

A jetliner's weight decreases as it consumes its fuel, but the wings cannot change size. Therefore, the wing loading decreases. If the plane remained at the same altitude, so that the air density remained constant, the decreasing wing loading would require the plane to slow down, thus affecting the schedule. Moreover, the efficiency of the jet engines would suffer, for they do not perform optimally at lower speeds. As its weight decreases, a plane must find more rarefied air in order to maintain its speed. If its wings have become too large for flying at an altitude of 10 kilometers, it moves up a kilometer. The engines don't mind: when the weight decreases, so does the drag if the finesse remains the same. Less drag means that the engines need to deliver less thrust. It does not matter, therefore, that as the cruising altitude increases the engines breathe somewhat thinner air.

On a long intercontinental flight, say from Tokyo to Amsterdam, a Boeing 747-400 might start out cruising at an altitude of 9000 meters (30,000 feet) and a weight of 380 tons. Thirteen hours later, when it begins its descent toward Amsterdam over Berlin, its weight has decreased by one-third, to 250 tons, and in the meantime it has climbed to an altitude where the air density is also 66 percent of its initial value: 12,100 meters (40,000 feet) (table 6). Each hour it must climb roughly 800 feet in order to compensate for the fuel burned.

The early production models of the Pratt & Whitney high-bypass-ratio fan-jet engines for the 747 had lots of "teething problems," which caused several months of delay for Pan American. The engines "flamed out" for no apparent reason, and at full power the fan casings bent a millimeter out of round, causing a 10 percent increase in fuel consumption. But the setbacks were overcome, and the current versions of the Pratt & Whitney, General Electric, and Rolls Royce engines for the 747 have proved so fantastically fit for their job that a variety of other airframes have been designed around them. Thus, Lockheed's Tristar, Douglas's DC-10, and Boeing's 757, 767, and 777 share a little of the 747's glory.

The main reason for the sustained popularity of these engines is their low fuel consumption. At cruising altitude, each engine consumes about 3000 liters per hour, and in return it produces 6 tons of thrust at an airspeed of 900 kilometers per hour. Force times speed equals power; in metric units we're talking about 60,000 newtons times 250 meters per second, or 15 megawatts (20,000 horsepower). The fuel consumption then computes at 0.15 liters per horsepower per hour. Your car, which needs about 20 horsepower at a speed of 100 kilometers per hour and uses about 7 liters per hour at that speed, consumes about 0.3 liters per horsepower per hour. This is in line with other numbers we have encountered. The thermal effi-

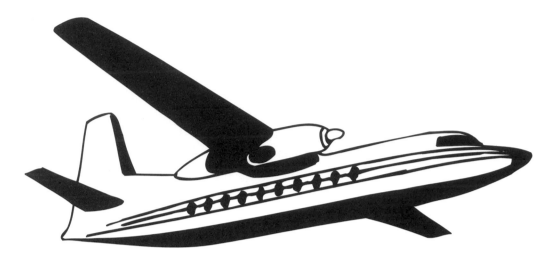

Fokker F-27 Friendship: $W = 2 \times 10^5$ N, $S = 70$ m^2, $b = 29$ m, $P = 2 \times 1600$ kW.

ciency of the piston engine in your car is only 25 percent, but a high-bypass-ratio jet engine, assisted by very low outside air temperatures and high cruising speeds, manages 50 percent! Both for the jet and for the car, the heat of combustion computes at 36 megajoules per liter (about 45 megajoules per kilogram), corresponding fairly well with the data given in chapter 2 (see table 3).

In a stationary application at sea level, a single engine of the type used in the 747 delivers about 20 megawatts—almost 30,000 horsepower. Because it is standing still and the air temperature is much higher than at 30,000 feet, the fuel consumption is much higher in this kind of application, although many other advantages remain. Because of their low weight, these engines are used to power some naval vessels. A 30,000-horsepower diesel engine weighs many tens of tons, but a comparable fan-jet engine weighs only 4 tons. Together, the four engines of a 747 account for less than 5 percent of the plane's takeoff weight.

The quick response of jet engines explains their popularity as auxiliary units in electric power generating plants. Just push a button and a few seconds later 20,000 kilowatts of emergency power can be delivered to the high-voltage grid. It takes hours to get up to full steam with a coal-burning or a gas-burning unit.

More than 2000 years ago the Romans invested heavily in a great highway system. The empire needed a solid road infrastructure in order to maintain unity and suppress local conflicts. The Roman highways were engineered so well that their remnants can still be found all over Europe. In seventeenth-century Holland, canals were dug to promote passenger traffic on a regular timetable, independent of fog and wind. A Haarlem banker could commute to Amsterdam, do his business, and be back home in time for dinner. The mobility of the merchant class helped to keep the United Provinces together. In the eighteenth century, England and France built extensive canal systems, too, judging investment in infrastructure to be the best preparation for the future. A century later, railroads put their webs across entire continents. The first transcontinental railroad in the United States dates from 1869, when the Union Pacific met the Central Pacific at Promontory Point in Utah. Nearly a century later, this infrastructure was supplemented by the interstate highway system, originally conceived, like its Roman predecessor, for rapid movement of military materiel. Over and over again, people have learned that adequate transportation facilities are necessary to keep a country united.

When Marshall McLuhan dreamed of the "global village," he was thinking primarily of telecommunications. He felt that the incessant chatter on intercontinental party lines would inevitably bring people closer together. But McLuhan did not pay much attention to the continuing need for physical mobility. Ultimately, phone and fax and even electronic communication are not enough. The time comes when you want to see the Grand Canyon or the Pyramids for yourself, meet and talk to people from other continents, or sip beer in the back yard of your international business partner. The 747 is the commuter train of the global village.

Appendix: Flight Data for Selected Birds

Birds are listed in order of ascending weight. The weight W is given in newtons, the wing area S in square meters, and the wingspan b in meters. The cruising speed V (meters per second) is computed with $W/S = 0.38\,V^2$. Female members of the *Falco* and *Buteo* families often are considerably larger than their mates; in that case, averages between the two sexes are given. Sources: Crawford H. Greenewalt, "Dimensional relationships for flying animals," in *Smithsonian Miscellaneous Collections,* no. 144 (1962); Colin J. Pennyquick, *Journal of Experimental Biology* 128 (1987): 335–347 and 150 (1990): 171–185; various field guides for birdwatchers.

	W	S	b	V
Black-chinned hummingbird *Archilochus alexandri*	0.026	0.0013	0.09	7.3
Ruby-throated hummingbird *Archilochus colubris*	0.030	0.0012	0.09	8.1
Goldcrest (Europe) *Regulus regulus*	0.040	0.0032	0.14	5.7
Golden-crowned kinglet *Regulus satrapa*	0.058	0.0051	0.17	5.5
Ruby-crowned kinglet *Corthylio calendula*	0.067	0.0058	0.18	5.5
American redstart *Setophaga ruticilla*	0.080	0.0063	0.18	5.8
Long-tailed tit (Europe) *Aegithalos candatus*	0.080	0.0060	0.18	5.9
Magnolia warbler *Dendroica magnolia*	0.092	0.0069	0.20	5.9
Winter wren *Nannus hiemalis*	0.094	0.0041	0.16	7.8
Crested tit (Europe) *Parus cristatus*	0.10	0.0073	0.20	6.0
Blue tit (Europe) *Parus caeruleus*	0.10	0.0066	0.21	6.3
Wren (Europe) *Troglodytus troglodytus*	0.10	0.0040	0.17	8.1

	W	S	b	V
House wren *Troglodytus aedon*	0.11	0.0048	0.17	7.8
Black-capped chickadee *Penthestes atricapillus*	0.12	0.0076	0.21	6.6
Redstart (Europe) *Phoenicurus phoenicurus*	0.13	0.0091	0.26	6.1
Reed warbler (Europe) *Acrocephalus scirpaceus*	0.13	0.0067	0.20	7.1
Sand martin (Europe) *Riparia riparia*	0.15	0.012	0.31	5.7
Rough-winged swallow *Stelgidopterix ruficollis*	0.16	0.011	0.30	6.2
House martin (Europe) *Delichon urbica*	0.16	0.010	0.29	6.5
Barn swallow *Hirundo erythrogaster*	0.17	0.012	0.31	6.1
Chimney swift *Chaetura pelagica*	0.17	0.010	0.31	6.7
Nightingale (Europe) *Luscinia megarhynchos*	0.17	0.010	0.25	6.7
Robin (Europe) *Erithacus rubecula*	0.18	0.009	0.23	7.3
Barn swallow (Europe) *Hirundo rustica*	0.20	0.013	0.33	6.4
Tree swallow *Tachycineta bicolor*	0.20	0.013	0.32	6.4
Great tit (Europe) *Parus major*	0.21	0.010	0.23	7.4
Song sparrow *Melospiza melodia*	0.22	0.009	0.23	8.0
White wagtail (Europe) *Motacilla alba*	0.22	0.013	0.28	6.7
Orchard oriole *Icterus spurius*	0.23	0.010	0.24	7.8
Tufted titmouse *Baeolophus bicolor*	0.23	0.012	0.25	7.1
Leach's petrel *Oceanodroma leucorhoa*	0.27	0.025	0.48	5.3
House sparrow *Passer domesticus*	0.28	0.010	0.23	8.6
Skylark (Europe) *Arlanda arvensis*	0.30	0.016	0.32	7.0
Swift (Europe) *Apus apus*	0.36	0.017	0.42	7.5

	W	S	b	V
Wilson's storm petrel *Oceanites oceanicus*	0.38	0.022	0.39	6.7
Purple martin *Progne subis*	0.43	0.019	0.41	7.7
Spotted sandpiper *Actitis macularia*	0.48	0.015	0.36	9.2
Common sandpiper *Tringa hypoleucus*	0.50	0.015	0.36	9.4
Golden oriole (Europe) *Oriolus oriolus*	0.72	0.027	0.47	8.4
American robin *Turdus migratorius*	0.82	0.024	0.38	9.5
Starling *Turdus vulgaris*	0.84	0.019	0.37	10.8
Blue jay *Cyanocitta cristata*	0.89	0.024	0.38	9.9
Blackbird (Europe) *Turdus merula*	0.90	0.025	0.40	9.7
American kestrel *Falco sparverius*	1.14	0.036	0.52	9.1
Common tern *Sterna hirundo*	1.20	0.053	0.82	7.7
Sharp-shinned hawk *Accipiter velox*	1.34	0.052	0.53	8.2
Common diving petrel *Pelecanoides urinatrix*	1.37	0.022	0.39	12.8
Merlin (Europe) *Falco columbarius*	1.45	0.044	0.60	9.3
Kestrel (Europe) *Falco tinnunculus*	2.00	0.068	0.74	8.8
Sparrow hawk (Europe) *Accipiter nisus*	2.50	0.080	0.75	9.1
Black-headed gull (Europe) *Larus ridibundus*	2.60	0.085	0.97	9.0
Barn owl (Europe) *Tyto alba*	2.80	0.120	1.00	7.8
Rock dove *Columba livia*	2.90	0.075	0.80	10.1
Moorhen (Europe) *Gallinula chloropus*	3.0	0.050	0.60	12.6
Black skimmer *Rhynchops niger*	3.0	0.089	0.99	9.4
Laughing gull *Larus atricilla*	3.25	0.106	1.03	9.0
Blue-winged teal *Querquedula discors*	3.32	0.037	0.61	15.4

	W	S	b	V
Little blue heron *Egretta caerulea*	3.40	0.134	0.98	8.2
Broad-winged hawk *Buteo platypterus*	3.76	0.101	0.84	9.9
Arctic skua *Stercorarius parasiticus*	3.90	0.117	1.05	9.4
Partridge *Perdix perdix*	4.0	0.043	0.53	15.6
Puffin *Fratercula arctica*	4.0	0.037	0.55	16.9
Cooper's hawk *Accipiter cooperi*	4.28	0.090	0.71	11.2
American coot *Fulica americana*	4.35	0.060	0.64	13.8
Royal tern *Sterna maxima*	4.70	0.108	1.15	10.7
Kittiwake *Rissa tridactyla*	4.9	0.097	1.05	11.5
Common gull (Europe) *Larus canus*	5.0	0.15	1.20	9.4
Barn owl *Tyto alba pratincola*	5.05	0.168	1.12	8.9
Marsh hawk *Circus hudsonius*	5.14	0.154	1.07	9.4
Ruffed grouse *Bonasa umbellus*	5.17	0.053	0.56	16.0
Shoveler *Spatula clypeata*	5.70	0.057	0.79	16.2
Razorbill *Alca torda*	6.20	0.046	0.66	18.8
Groshawk (Europe) *Accipiter gentilis*	7.0	0.26	1.25	8.4
Marsh harrier (Europe) *Circus aeruginosus*	7.0	0.22	1.35	9.1
Peregrine falcon (Europe) *Falco peregrinus*	8.0	0.13	1.05	12.7
Red-shouldered hawk *Buteo lineatus*	8.0	0.166	1.02	11.3
Fulmar *Fulmarus glacialis*	8.1	0.124	1.13	13.1
Great egret *Casmerodius albus*	8.7	0.222	1.34	10.2
White ibis *Eudocimus albus*	9.0	0.160	0.95	12.2
Herring gull *Larus argentatus*	9.5	0.203	1.36	11.1

	W	S	b	V
Guillemot *Uria aalge*	9.5	0.054	0.71	21.5
Duck hawk *Falco peregrinus anatum*	9.7	0.124	1.02	14.3
Pintail *Dafila acuta*	9.7	0.076	0.89	18.3
Buzzard (Europe) *Buteo buteo*	10	0.27	1.35	9.9
Red-tailed hawk *Buteo borealis*	10.9	0.209	1.22	11.7
Eastern goshawk *Astur atricapillus*	11.1	0.174	1.11	13.0
Pheasant *Phasianus colchius*	12	0.088	0.72	18.9
Brent goose (Europe) *Branta bernicla*	13	0.14	1.20	15.6
Roseate spoonbill *Ajaia Ajaja*	13	0.226	1.25	12.3
Mallard *Anas platyrhynchos*	13.2	0.099	0.90	18.7
Great skua *Catharacta skua*	13.5	0.214	1.37	12.9
Blue heron (Europe) *Ardea cinerea*	14	0.36	1.70	10.1
Double-crested cormorant *Phalacrocorax auritus*	14.1	0.179	1.16	14.4
Osprey *Pandion haliaetus*	14.9	0.300	1.59	11.4
Turkey vulture *Cathartes aura*	15.5	0.44	1.75	9.6
Spoonbill (Europe) *Platalea leucorodia*	16	0.25	1.40	13.0
White-fronted goose *Anser albifrons*	17	0.18	1.40	15.8
Shag *Phalacrocorax aristotelis*	18.1	0.158	1.04	17.4
Great black-backed gull *Larus marinus*	19	0.27	1.73	13.6
Great blue heron *Ardea herodias*	19.2	0.419	1.76	11.0
Black vulture *Coragyps atratus*	20.8	0.327	1.38	12.9
Cormorant (Europe) *Phalacrocorax carbo*	21	0.25	1.60	14.9
Common loon *Gavia immer*	24.3	0.136	1.47	21.7

	W	S	b	V
Great white heron *Ardea occidentalis*	25	0.493	1.91	11.5
Sooty albatross *Phoebetria palpebrata*	28.4	0.338	2.18	14.9
Gannet *Sula bassana*	30.1	0.262	1.85	17.4
Greylag goose *Anser anser*	31	0.27	1.63	17.4
Brown pelican *Pelecanus occidentalis*	33.9	0.450	2.26	14.1
White stork (Europe) *Ciconia alba*	34	0.50	1.98	13.4
Golden eagle (Europe) *Aquila chrysaetus*	37	0.54	2.10	13.4
Black-browed albatross *Diomeda melanophris*	37.9	0.356	2.16	16.7
Grey-headed albatross *Diomeda chrysostoma*	37.9	0.352	2.18	16.8
Golden eagle *Aquila chrysaetus c.*	46.6	0.65	2.00	13.7
Bald eagle *Haliaetus leucocephalus*	46.8	0.756	2.24	12.8
Giant petrel *Macronectus giganteus*	51.9	0.331	1.99	20.3
Canada goose *Branta canadensis*	56.6	0.282	1.70	23.0
Whistling swan *Cygnus columbianus*	59.4	0.416	2.15	19.4
Griffon vulture (Europe) *Gyps fulvus*	75	1.00	2.60	14.0
Wandering albatross *Diomeda exulans*	87.3	0.611	3.03	19.4
Mute swan *Sthenelides olor*	116	0.681	2.40	21.2

Bibliography

R. M. Alexander, *Locomotion of Animals.* Blackie, 1982.

J. D. Anderson, *Introduction to Flight.* McGraw-Hill, 1989.

S. Childress, *Mechanics of Swimming and Flying.* Cambridge University Press, 1981.

N. Elkins, *Weather and Bird Behavior.* Poyser, 1983.

C. H. Greenewalt, "Dimensional relationships for flying animals." In *Smithsonian Miscellaneous Collections,* no. 144 (1962).

H. Hertel, *Structure, Form, Movement.* Reinhold, 1966.

C. Irving, *Wide-Body: The Making of the 747.* Hodder & Staughton, 1993.

B. W. McCormick, *Aerodynamics, Aeronautics and Flight Mechanics.* Wiley, 1979.

R. von Mises, *Theory of Flight.* Dover, 1959.

U. M. Norberg, *Vertebrate Flight.* Springer-Verlag, 1989.

J. C. Pennyquick, *Animal Flight.* Edward Arnold, 1972.

J. C. Pennyquick, *Bird Flight Performance.* Oxford University Press, 1989.

G. J. J. Ruygrok, *Elements of Airplane Performance.* Delft University Press, 1990.

G. Rueppell, *Bird Flight.* Van Nostrand Reinhold, 1977.

R. S. Shevell, *Fundamentals of Flight.* Prentice-Hall, 1983.

B. Stillson, *Wings: Insects, Birds, Men.* Gollancz, 1955.

J. W. R. Taylor, *Jane's All the World's Aircraft.* Jane's Information Group, 1988.

E. Torenbeek, *Synthesis of Subsonic Airplane Design.* Delft University Press, 1982.

V. A. Tucker, "Respiratory exchange and evaporative water loss in a flying budgerigar." *Journal of Experimental Biology* 48 (1968): 67–87.

Index